WITHDRAWN
UTSA LIBRARIES

SMART CARDS
The New Bank Cards

SMART CARDS
The New Bank Cards

Updated and Expanded Edition

JEROME SVIGALS

MACMILLAN PUBLISHING COMPANY
NEW YORK
Collier Macmillan Publishers
LONDON

Published 1985, 1987 by Macmillan Publishing Company.
Published 1985. Revised Edition 1987

Macmillan Publishing Company
866 Third Avenue, New York, NY 10022

Collier Macmillan Canada, Inc.

Printed in the United States of America

Printing: 1 2 3 4 5 6 7 8 9 10

Year: 7 8 9 0 1 2 3

Library of Congress Cataloging-in-Publication Data

Svigals, Jerome.
Smart cards.

Includes index.
1. Smart cards. I. Title.
TK7895.S62S85 1987 332.1'78 87-13745
ISBN 0-02-948901-6

To Our Grandchildren

Sarah, Logan and Brian

May peace permit your successful pursuit of an exciting future

Contents

Preface

Early Power Plants

The start of the industrial revolution was based on large central power plants. First, it was the waterwheel. Power was distributed from the waterwheel by a family of shafts and belts. Waterwheels, as an energy source, were later replaced by steam engines and then electric motors. However, similar physical energy distribution systems were used from the large central source.

As with most technologies, evolution took the form of economic abundance. Grandfather clocks evolved to wrist watches. Quill pens became ballpoint pens. Horsepower evolved to compact energy sources. In time, the fractional horsepower electric motor appeared. It helped to accelerate factory productivity. For the first time, the fractional horsepower motor could be easily used to economically match power needs. The motors could be used where needed and with the amount of power required for specific applications. By World War II the fractional horsepower motor function and economics had reached a level that justified its wide-scale home use. Today, an average home has over 100 electric motors: clocks, refrigerators, air conditioners, heaters, washers, driers, sewing machines, vacuum cleaners, thermostats.

To a large extent, we are going through a similar cycle of events with information processing systems. Shortly after World War II, the large vacuum-tube-based digital computer system was introduced.

Within a period of twenty years, the technology evolved rapidly from vacuum tubes to transistors to integrated circuit chips for large systems. However, as this digital technology evolved, it also started to offer economic "fractional horsepower" computers. In fact, it has evolved so rapidly that today we have arrived at the era of the fractional computer in the form of a microcomputer-on-a-chip.

The Chip

What is the integrated circuit chip? It is high-density electronic circuits "painted" on a base of silicon by electrochemical means. The change in density of electronic circuits as a result of the last thirty years of technological evolution is unbelievable. In fact, the first electronic digital computer in 1950 was two million times the size of today's microcomputer-on-a-chip! (1000 square feet versus a one-quarter-inch square.) The microcomputer-on-a-chip is faster, has more capacity, and is more flexible than the original tube-based computer.

Today's household uses perhaps 20 to 30 chip-based microprocessors. They are in microwave units, portable radios, TV sets, hi-fi's, and family cars. Just as the fractional horsepower motor was poised for a population explosion forty years ago, so is the integrated circuit chip today. This book is the story of one part of the next twenty years. It is an educated projection of the role to be played by the chip in our everyday life. It is an attempt to forecast how the microprocessor-on-a-chip will be packaged, personalized, tailored, and used by the whole population.

There have been a number of products that package several microcomputers-on-a-chip in the form of a small ("personal") computer. However, history shows that all great inventions go through this early stage. The new technology is used first to provide the "old" solution, like the iron horse and the horseless carriage. Today's personal computer is a modern example. It demonstrates two transient concepts. The first is that a knowledge of computer languages and programming is necessary to effectively use information processing capabilities. The second is that the best way to create information processing users is to train them as "mechanics" of the keyboard and of computer interface languages. Current personal computer offerings based on these concepts are not encouraging the broad general usage of information processing power. If the modern TV set had the user-operating complexity of today's personal computer, would we have seen the same level of TV set usage by the general public?

Technology Changes Our Lives

Fortunately, we have some excellent examples of a more mature development of new technologies. They demonstrate that a new technology will change our lifestyle when the human interface to the technology is simplified to allow direct use in everyday life. One example is the automobile. Today the automobile has created the shopping center and supermarket based on the mobility of the family car. Effective use of the family car doesn't require an understanding of internal combustion engines or the proper sequence of using a carburetor. Similarly, the automatic teller machine (ATM) has achieved wide public use of a computer system in a box. The automatic teller machine is changing our lifestyle while conditioning our society to 24-hour and 7-day banking. This broad acceptance is an adaptation of sophisticated technology to untrained public users.

In both of these cases the service and product developer has mastered a basic goal. Namely, the ability of the public to use new technology without a knowledge of the technology. This book is about the same concept of use for the microcomputer-on-a-chip. This concept is called the SMART CARD. The Smart Card packaging, its economic application, and the requisite user skills will be described. It is the successful development and application of factors that will contribute to the successful use of many types of microcomputers-on-a-chip Smart Cards in our everyday life.

The Chip and Tomorrow

The chip-based devices of tomorrow will be even more dense, more economic, with a higher performance, and very adaptable. But how will we mold tomorrow's abundance of chip-based information processing power? Interestingly the chip abundance arrives in a period of shortages: of energy, government funds, employment, leisure time, investment in education, third world economic growth, and productivity levels. The abundance and broad presence of microcomputer chips will have a remarkable impact on these issues. The societal impact will be discussed later—after we describe tomorrow's Smart Card—the Ultimate Personal Computer.

The Smart Card is *not* only a means of making financial payments. However, many of the Smart Card concepts have come from two decades of electronic and self-service banking development. The successful introduction and customer acceptance of these new banking facilities has been an important school. ATM users have been observed while using the machine, bankers have devoted many hours to their product development and market introduction, and equipment development professionals have acquired years of evolving experience. This book additionally benefits from more than a decade of development by a number of innovative Smart Card inventors. To all, I am indebted for the concepts contained herein. To each I extend my appreciation.

A special acknowledgement and expression of appreciation is offered for the reviews, comments, and contributions of Dr. Stephen Weinstein, Chief Scientist, The American Express Company. Dr. Weinstein also wrote an important Smart Card article published in the *IEEE Spectrum* (February 1984).

There is now a stream of Smart Cards, accepting products and devices, new applications, international standards, and new economic justification models. More important, there is now recognition at the national government level (e.g., Japan and France) that Smart Cards justify supporting government policies and actions. This comes from the observation that hand-held and used, integrated circuit chip packages (Smart Cards) will make a contribution to the positive evolution of our society. It is reassuring that both the societal profits and societal improvements are now starting to occur. Much more is anticipated.

Woodside, California, 1987

SMART CARDS
The New Bank Cards

1 Introduction

WHAT IS A SMART CARD?

The Smart Card consists of an integrated circuit chip or chips packaged in a convenient form to be carried on one's person. The package may be a bank card with internal integrated circuit chips or it may be a replacement for a soldier's dog tag. It may be almost any shape, size, and thickness that is convenient for the user to carry, insert, or connect into a device and retrieve. Smart Card packages range from the simplest of units to a work station, which may include its own keyboard, display, power, and communications interface. In fact, it might be a hand-held, portable computer to which the appropriate microchip has been added for Smart Card functions.

The Smart Card chip will have several functional abilities, possibly including logic. This is the ability to provide and follow instructions, to make logical choices, and to follow alternative decision paths. Logic is also used to recognize and respond to externally provided information. The logic process may be executed internally in the chip or logic results may be provided to the Smart Card externally. The chip may be general or special purpose designs. The chip may include a complete microprocessor or they may be specialized processing chips such as an encryption device. In short, any and all of the logical capacities developed for microcomputer-on-a-chip technology may be used in the Smart Card.

Smart Card chips may have temporary or permanent data storage functional ability. The storage content may be read externally off the Smart Card or it may be used internally for information processing and decision making. The chip may also have destroyable or irreversible, alterable entry or storage circuits. For example, data may be written into a storage area and the write circuits are then destroyed (perhaps by burnable fuse links). This will prevent alteration of the written data. The Smart Card may also be designed to selectively destroy stored data. This function would "use" purchase increments as they are consumed, as in a prepaid telephone multiple-call units card.

WHY THE SMART CARD?

The power and capacity of integrated circuit chips is growing rapidly. Consider the following projections:

YEAR	CAPACITY PER ¼" CHIP (no. of elements)
1970	10,000
1980	150,000
1990	1,000,000
2000	2 to 4,000,000

Chip costs are also falling rapidly, perhaps 20 percent to 40 percent per year for equivalent capacity or function. Chip packaging in plastic, technological alternatives, function versus density trade-offs, and physical densities are all evolving at a rapid pace. (The future price evolution of Smart Cards will be discussed later in Chapter 12.) The combination of card function and chip capability opens a vast new marketplace paced by the development of economically viable applications. The market sales volume of microcomputers-on-a-chip is becoming the dominant economic factor in the information processing industry. These chips include both logic and storage capabilities. Table 1.1 shows the dollar value distribution of complete products. It shows, for example, that microcomputers will grow to 55 percent of a $313 billion market by the year 1995.

After the mid-1980s, the microcomputer-on-a-chip will be the dominant form of computers. This will occur while the total computer industry grows by a factor of 24. The entry year into the marketplace of a microcomputer-on-a-chip depends on the information width handled

TABLE 1.1

YEAR	TOTAL MARKET ($B)	MAIN FRAMES (%)	MINI-COMPUTER (%)	MICRO-COMPUTER (%)	OTHER (%)
1975	13	83	10	0	7
1980	29	60	17	6	17
1985	63	36	21	20	23
1990	141	17	21	37	25
1995	313	6	16	55	23

by the chip. This width is measured in the number of bits (a "0" or a "1" of information).

YEAR OF INTRODUCTION	MICROCOMPUTER-ON-A-CHIP (WIDTH IN BITS)
1978	16
1985	32
1992	64

How will the microcomputer-on-a-chip be packaged? A large portion will be put *into* and *under* the covers of other products. But how will their power be tapped by the largest set of users—the public? This question is the focal point of the Smart Card opportunity. Will the microcomputer-on-a-chip be directly usable by the public? The answer is affirmative, given the right package, application, and usage structure. The ability to package chips economically is now being demonstrated in common applications, ranging from wrist watches to hand calculators. These are carefully prespecified uses of chips. Although multi-functioned, the public enjoys these uses of chip technology. What about more general applications?

THE SMART CARD OPPORTUNITY

A big opportunity lies in tapping the capacity of each Smart Card to contain a tailored application. The Smart Card will be used to provide a much less expensive and much easier alternative to using a conventional computer. Chips are reaching a level of information processing capability matching that of a complete personal computer, except for the peripheral devices. The key difference is in the tailored chip design versus the general purpose personal computer. The tailored design allows significant differences that compress the logic and storage needed, simplify the user interaction, and speed the responses possible with the Smart Card. (More on those important differences later.) Hence, with the right application and usage structure, it will be less expensive to provide and use several Smart Cards. Each of these Smart Cards, when used with an appropriate information processing appliance, will compete economically with a general purpose computer or work station with its prorated support and dedicated line costs. Each Smart Card, which might cost a few dollars, would contain the equivalent of a fully tailored information processing and application capability.

How does a Smart Card with one or a few chips have the capacity of a personal computer? It doesn't! It has an information content that allows supervision, control, and interaction with an associated device or system. By prestructuring specific applications and tailoring the design and the system responses, the user-carried Smart Card—with a minimum choice of added options or variables—allows the entire operation, no matter how complex, to appear simple to the user. A good example is the magnetic-strip bank card, which controls an automatic teller machine (ATM) with a minimum of user knowledge of the internal machine operation.

Individual Smart Cards will be considerably easier than personal computers for the public to use. No hunt-and-peck keyboards. No computer languages or procedures. No programs to write, no complex display interaction. Not even a need to understand program loading and access procedures. Smart Cards will achieve the level of general acceptance and use enjoyed by the television set. They will probably achieve public usage equal to that of the telephone direct dial system. In fact, Smart Cards may find their most important use with telephone and television set usage. But more on that later!

WHY IS THE SMART CARD BETTER?

Current systems based on magnetic-stripe, embossed plastic cards have limitations. They have limited storage capacity. They are passive devices without built-in logic for security control of the content or access to read or change the content. Stripes require reading techniques that are more mechanical and serial in nature. The Smart Card brings, according to its advocates, more capacity, better control, easier connection, and direct electronic readability of the integrated circuit chips. These claims will be discussed in depth in later chapters.

THE SMART CARD SOCIETY

Imagine that we are in the Smart Card Society. We each have several Smart Cards, drawn from the following list.

SMART CARD TYPE	ISSUER
Financial services	Financial institutions
Medical profile and services (profile from your doctor)	Hospital insurance
Government identification (Licenses from appropriate dept.)	Local, state, and national Governments
Communications services	Communications service co.
Travel services card	Travel services co.
Employment access and reporting	Employer
Military skills and training	Military
Electronic diagnosis	Electronic device vendor
Automobile routing	Automobile association
Work station personalization	Employer, computer servicer, terminal, and cable services
Software loading and protection	Developer or vendor

INFORMATION PROTOCOLS

Each Smart Card is issued by an "identifiable" organization, which has received identifying code assignments from a national registration authority. Each card has a standard data content or *information protocol*.

The protocols provide for the following application transaction functions:

Identification of the card issuer and holder.
Initiation of entry to the card content by a holder personal identification number (PIN).
Setting available services and their limits for this card holder.
Establishing the control dates for this card: to start, to renew allowances, and stop use.
Accessing reference information, e.g., telephone and network access addresses.
Capturing Smart Card use in an internal transaction journal.
Containing personal data base, e.g., financial data and access codes.

The information protocols are the basis for broad mobility of Smart Card usage. For example, assume that every telephone or telephone-related device (Videotex, electronic directory, Local Area Data Transport long-distance facility entry point) has a Smart Card interface and information protocol logic. You, the Smart Card user, could go to any device, in any location, and carry out a wide range of Smart Card-based transaction activity. The Smart Card society will offer a card interface or plug-in option in many locations. The interface is in many *information appliances* such as telephones, television sets, videotex units, a retail store checkout counter, a hotel registration desk, a car rental desk, your doctor's desk, your automobile, your electronic mail receiver, your home diet unit, a satellite broadcast descrambler, a stockmarket transaction unit in your office building lobby, and an electronic magazine receiving unit at home.

USING THE SMART CARD

In each application the common information protocols allow the Smart Card-based operation sequence to be the same.

Insert the user's Smart Card.
Establish your personal identification by Personal Identification Number entry, validated by comparison inside the Smart Card.
Select from a menu of options.
Provide a minimum data entry to complete the transaction request.
Confirm or accept the proposed action and amounts.
View the result, make a final decision.
Remove any receipts.
Remove the Smart Card with a record of the decision or action taken.

This sequence of operations will be the fundamental approach of the Smart Card era. The general population is being conditioned to this very workable sequence by today's automatic teller machines. Thus, introduction of the Smart Card for a financial transaction card or other applications is a compatible and evolutionary step. This can be expected to encourage broader use of the Smart Cards.

TYPES OF SMART CARDS

In the Smart Card society there will be at least three types of Smart Cards. First, a user-carried card that contains logic, data, and responses for the information protocols described above. Second, a *job* or *provider* card that contains the structured details, data, and communication addresses required for the tailored task or job. Third, a *device* card that, if needed, operates the *information processing appliance* (the Infopro) needed to perform the desired task.

Not all card types are needed all of the time. In some cases, only the user card will be needed, such as at an automatic teller machine. The job or device card function would then be built-in to the information appliance in the form of extended storage and logic capacity. In a retail point-of-sale unit, the job card will also be used. It contains application and security logic and a list of "hot" cards to be used instead of making an authorization phone call. The job card will also capture the transaction details. The merchant will deliver a batch of captured transaction data on his job card to his financial institution by phone or physical carriage. In an Infropro, the device card will provide configuration control. This allows the same set of building blocks (e.g., display, input/output, communications interface), to be used in a variety of ways, depending on the application needs.

THE INFORMATION APPLIANCE

The information processing appliance—the Infopro—will be a significant outgrowth of Smart Card use. Consider the Infopro in your home work area or office. It will have a flat TV screen with a touch-entry facility instead of a keyboard, and an automatic communications device, a printer, and a modest capacity storage device. The Infopro will have insert positions for up to three Smart Cards. Consider a financial application:

First, insert your financial transaction user card.

Second, select and insert the Cash Flow Management application or job card. This card includes the telephone number of your financial data base and supporting analysis program. It also includes the encryption key for protecting vital data between your unit and the financial institution.

Third, insert the device card which specifies and controls the work station or Infopro configuration for this application. This includes a touch entry display, a communications unit, a local storage, an output printer, and a local logic unit.

A SMART CARD EXAMPLE: PERSONAL FINANCIAL MANAGEMENT

After insertion of the Smart Card, which triggers an automatic preparation, a menu of options appears on the display as follows:

1. Transactions and bills
2. Statements and inquiries
3. Cash flow analysis
4. Financial decisions and management

You select "Transactions and bills." The latest bills received through electronic mail appear. For each you confirm or modify the amount to be paid and the date of payment. A bill from a new biller appears. You confirm its entry into your bill payment system. The new electronic bill contains the necessary data for account identification, critical payment date options, and the electronic mail address for payments.

Next you select "Cash flow analysis." It indicates the changes in cash flow since your last use of this application. Cash in, cash out, and total transactions are shown. It provides a cash flow analysis against the experiences of the last 12 months of cash flow. Are you cash short or do you have an excess? You have some added cash. You examine another option menu and take several actions. First, you update the cash, credit line, and debit content of your financial transaction card. Second, you identify a small amount of excess cash that would normally be swept into a money fund. You decide to consider other alternatives.

Finally you select "Financial decisions and management." You enter the amount and the time period for the available funds. A list of current alternatives, their expected return, and other constraints are shown. You select and enter your investment decision including amount confirmation, action dates, and the limit conditions and actions. You remove the job and device cards. You have no further application needs now. However, when the unit is not otherwise in use you elect to leave it in an electronic mail receiving mode. You enter your electronic mail receiving job Smart Card and the appropriate device configuration card. You leave the unit for unattended operation.

SMART CARD COMPARED TO THE PERSONAL COMPUTER

I think you have the idea. Smart Cards are going to bring the power of information processing to everyday basic life tasks and decisions. More specific information and more decision processes will become routinely available. The choice of appropriate cards from a library of job and device cards will offer a broad spectrum of application power and facility. The era of having to learn programming as the key to making use of computer power will end. In each case, the application, job and device knowledge will be built into the microcomputer-on-a-chip and handled as Smart Cards. Through simple menus, option tables, and action confirmation, the information appliances will harness the full power of the information processing age.

How do we know this is possible? By considering the excellent experience with conventional magnetic-stripe-card-based system devices, such as an automatic teller machine. Excellent results have been achieved with massive public usage on a worldwide basis, and are worth considering as a guide to future Smart Card development.

LESSONS FROM THE ATMs

The automatic teller machine (ATM) is a complex information processing system. It is a complete computer system in a box. It has up to ten input and output devices. It handles at least four media including currency, cards, receipts, and envelopes. It communicates to remove computers. It has self-supervising operating, application, and diagnostic programs. It incorporates sophisticated physical and logical security features. Its operation is structured, with all transactions authorized and controlled.

Yet, operation of the ATM by the general public has been reduced to a very simple sequence. Young children can operate it successfully. Retired persons enjoy its convenience and services.

A typical cash dispensing action takes as little as 15 to 20 seconds. It requires as few as four input actions for a complete and successful use. Insert your financial transaction identifying card, enter your personal identification number (PIN), select the transaction, and specify or confirm the amount requested. All other actions, communications, and controls are automatically performed by the prestructured logic, data, and communications facilities in the ATM.

Don't be mislead into thinking that these are expensive results. Today's ATM includes expensive physical security and media handling capabilities that would be avoided in a "no money—no deposit" unit. The price includes around-the-clock availability in a wide range of external environmental conditions. When the Smart Card becomes available, we could use it as a cash equivalent. At that time, the need to dispense cash and accept its protection costs in an ATM would start to disappear. Thus, the Smart Card could help reduce the complexity and cost of ATMs. Public acceptance of the Smart Card cash alternative would allow cash to be obtained through any Smart Card communicating information appliance. We will say more later about ATM experience and their evolution into the Smart Card era.

THE PURPOSE OF THIS BOOK

This book is devoted to the many interesting facets of the Smart Card. This book is nontechnical and intended to help Smart Card service providers, card suppliers, and issuers as well as its potential users. It describes the vital role of the Smart Card in leading us into a true information processing society, a society in which almost all members will enjoy Smart Card advantages. Of special importance are the lessons learned from public use of earlier successful solutions such as the Automatic Teller Machine. That experience will help answer the vital questions centered on human factors—security and economic justification.

2 Smart Cards for Financial Transactions

A leading contender for early Smart Card introduction is the bank card application area. There are about 350 million magnetic-stripe bank cards issued in the United States today. There may be up to twice that number of magnetic-stripe cards issued worldwide. In the card standards area, these are called financial transaction cards (FTC). Since most people are familiar with bank cards, I will use them as examples in this quick application view.

THE FINANCIAL TRANSACTION CARD

Data storage and handling is a function of existing bank cards. They provide the following functions:

1. Human-readable data: printing on the card identifies the card issuer, usable time periods, the customer's name and account number, the customer's signature, and other pertinent data. This data is the foundation for use of the financial transaction card as a payment device in routine financial transactions.
2. Machine-readable data: the raised embossing on the bank card is used for imprinting a paper form. This provides paper receipts to the merchant and the customer with transaction details. It also

provides a machine readable document to be read in an optical reading machine. This reader captures the data for account updating and billing.

3. In addition, a magnetic stripe on the back of the card is used to provide readable identification for authorization and data capture systems. The card-reading device, e.g., an ATM, adds transaction details to create a data capture record without the need to process paper.

TODAY'S ATM TRANSACTION

In the ATM, the magnetic-stripe data provides information such as the issuer and customer account identification. This is the basis for an automatic authorization message, achieved by a direct communications link to a control point. The control point is usually the card issuer's host computer and its account data base. The card number is checked against reported lost and stolen cards. It is also checked against the individual account record for available funds. The transaction amount is used to update the available funds for further use of the card. In addition, several sensitivity tests are applied. For example, how many times has the card been used today? How many dollars have been requested today in total? These tests are used to detect certain types of problems, such as an unreported stolen card.

In ATM usage, there are several types of cards and account relations. A *debit* card gives access to a savings or checking account balance. A *credit* card gives access to a pre-established line-of-credit funds. Cards may be issued for use in the ATMs of the issuing bank only and are referred to as ATM *access* or *proprietary* cards. Debit and credit cards, such as Visa or Mastercard, may be issued for a national interchange.

In addition to the financial transaction card data (credit, debit, proprietary or interchange), a personal identification number (PIN) is requested in most unattended units. This PIN is also used to check that the card presenter is the owner of the card. The PIN has a number of sensitive protection issues. For example, if the PIN is transmitted in the clear on a communications line, then it may be easily read by a wire tapper. Hence, this type of sensitive data is encrypted. This requires that an encryption algorithm be provided at the entry point. That, in turn, requires that a "key" which personalizes the encryption process to a specific bank be sent to the encrypting unit. That, in turn, requires protection and "management" of the key at the PIN receiving point. All of these security steps take added equipment and precautions.

TOMORROW'S SMART CARD TRANSACTION

The Smart Card implementation of a similar process at a point-of-sale with a financial transaction card requires the following cards.

The User's Card

The customer's financial transaction card (the *user's* card) appears as a conventional bank card. The card is embossed and has a magnetic stripe. In addition, it has an integrated circuit chip within the card, communicated to by a set of contacts on the front or rear left center of the card surface. The contacts are above the embossing area and below the magnetic stripe. The issuer chooses the rear (eight contacts) or the front (eight contacts). The card does not have its own power. The card is ordinarily provided with power by the work station, although cards with their own power have been proposed.

The user's card has four types of storage areas:

1. Open or nonconfidential: This area contains public information, such as the card issuer identification and the account number of the card holder.
2. Confidential: This area contains sensitive data, such as the available account balances and permissible services for card use and their limits. This storage area may be accessed only after the card has been "opened" for a transaction. This occurs after the correct PIN number has been delivered to the card.
3. Inaccessible: This area contains data that is never readable or releasable outside of the card. The contents of this storage area is used within the card to make decisions. For example, it contains the PIN value acceptable for this card. The acceptable PIN value was entered by a personalization process at the time of card issue. This area may also include the encryption key value to be used with this card to protect vital transaction data.
4. Permanent recording area: This area is used to capture the unique details of each transaction and is called a *journal*. It is used to answer inquiries on prior card use and to recreate the transaction stream that has reduced the card-based account balance.

In addition to data storage, each user card contains logic. The following types of logic are provided:

1. Security: This logic controls access to the card content. It may also include encryption logic for protection of data to be transmitted.
2. Application: The rules and processes for creating and checking balances, updating amounts and generating receipts and journal entries are kept here. Also, the logic and controls for interacting with the card holder during the transaction are here.
3. Operation: Instructions that communicate with the job card and the communications line are here. This might include telephone numbers, exception transaction definition (e.g., prescribed dollar limits) and handling, error recovery and restart, setting today's date and time, and similar guidance.

The Job Card

The merchant or *job* card is also required. This card provides for additional storage and logic in a portable form. The merchant card has a larger capacity. It has to capture details of all merchant transactions for delivery to the merchant's bank. It generally has its own power. The content is generally erasable and the storage is rewritable. The storage areas of the job card contain the following:

1. Data capture: This area captures transaction details for later delivery to the merchant's bank for conversion to deposit dollars for the merchant. This data is then routed by the merchant's bank through the automatic clearing house network to the card issuer's bank. The captured data is used by the card issuer's bank for account billing and control. The captured data becomes part of a journal of entries.
2. Hot list: A list of reported lost, stolen, and over-limit account numbers are recorded by the bank each time the transaction details are delivered to the bank. These are hot list accounts for which transactions may be accepted only after a direct call to the bank for further authorization.
3. Confidential and control data: The encryption key to be used in the data protection process is stored here. It is used to generate check numbers to prevent journal alteration, additions, and deletions. Also, telephone numbers for the authorization backup may be here. These are used for over-limit transactions that the merchant would like to complete because of their high value. However,

these transactions must be authorized by an automatic dial-up call to the merchant's bank.

In addition, the merchant's job card contains logic. It participates in the logic process with the customer's user card and the work station device card logic. The merchant card logic includes:

1. Processing: Guidance for error checking and correcting. Logic for device guidance such as receipt formatting and data content.
2. Security: To participate in the access and information protection processes.
3. Operation: Instructions to assist in the applications part of the work station (e.g., point-of-sale) device, to display and print, to use the communications facility, and to control the transaction performance handling process.

The Device Card

The customer user card and the merchant job card are activated by their insertion into the work station. The work station also has logic and storage capabilities in its device card. It may be a portable device or a built-in functional ability. The work station has a set of processing modules necessary for transaction handling, including the following:

1. A display PIN pad module with customer Smart Card insertion facility: This unit is usually at the end of a cable for ease of customer use. A shield is provided to hide the display content and the customer's key entries.
2. A work station with merchant keyboard, display, telephone interface unit, merchant job card insertion device, receipt printer, and paper supply. The unit contains a power supply that also supplies power to the customer's Smart Card. The telephone line is connected to the telephone interface device. The work station provides for the device card or has the equivalent wired-in storage and logic.

THE TRANSACTION PIECES

At this point all of the pieces are in place. Let's go through a routine point-of-sale operation. The components are as follows:

Cards: Customer user, merchant job, and possibly device cards.
Logic: For application, security and operation. Distributed among the customer card, the merchant's card, and the work station or device card.
Data Storage: Distributed among the customer's user card, the merchant's job card, and the work station.
Communications interface: In the work station.
Customer interface unit: The display PIN pad with user Smart Card receptacle.
Merchant interface unit: The work station.
Transaction data capture: Customer's user Smart Card journal entry, merchant's job card transaction capture, and the printed receipt for both.
Authorization control: The customer's user card for available balances and allowable services, the merchant job card for the hot list and authorization limits and rules, the work station for today's date and limit controls, and an automatic telephone call for exception condition authorization (e.g., over-prescribed dollar limits).

SIMPLICITY GUIDE LINES

If all this sounds a bit complex, remember our objective. Our objective is to set the stage for *simplicity* of operation as seen by the customer and the merchant. A description of the inside of the public telephone network or the design of the television set circuits or the inside logic of a hand calculator would probably be even more complicated. In the use of the Smart Card simplicity is obtained, as in the ATM, by the following guide lines:

Leave the data capture and information processing complexities to the preprogrammed microcomputers-on-a-chip and the work stations.
Reduce the keyboard entry and display interaction with the user to a minimum.
Provide all other actions and condition handling in the built-in logic so as not to require either customer or merchant knowledge or training.

THE SMART CARD TRANSACTION

Having set the stage, let's start the play. A purchase with a Smart Card would seem appropriate. The sequence of events is as follows:

The customer comes to the check-out stand with an article to purchase and a Smart Card financial transaction card (FTC).

The customer inserts the FTC into the display PIN pad.

The merchant enters the transaction amount into the work station keyboard.

The transaction amount appears on the display PIN pad.

The customer depresses a key on the display PIN pad that accepts the transaction amount.

The guidance light on the PIN pad directs the customer to enter his PIN.

The entered PIN is sent to the customer's FTC for acceptance.

When the PIN is accepted by the FTC, it is opened for the transaction.

The transaction amount is sent to the Smart Card FTC. Are there enough funds? If so, the new balances are computed and the transaction is entered into the transaction journal of the FTC.

The transaction acceptance is sent to the work station. A receipt is printed for the customer and the merchant. Data are captured for the merchant job Smart Card transaction record. Secure check numbers are computed and added to the journal records to assure security in their further handling.

The work station and PIN pad guidance lights indicate transaction completion and request user Smart Card removal. The transaction is complete.

The Customer's View

The actual activity of action by the customer in this process is minimal, as follows:

Insert his user Smart Card FTC.

Follow the guidance lights.

Use the correct amount key.

Enter the PIN value.

Remove the Smart Card FTC.

Take the receipt and merchandise.

The Merchant's View

The merchant activities are also minimal, as follows:

Insert merchant job Smart Card and data at the beginning of the day.
For each transaction, key in the transaction amount and transaction type identification.
Follow the guidance lights.
Give the customer a copy of the receipt.
Keep a copy of the receipt in the cash drawer.

At the end of the day the merchant can have the transaction batch communicated to the bank. Or the merchant can take the job Smart Card to the local bank branch. At the branch the teller puts the job Smart Card into a work station, which reads the card content. The transaction data content is read and captured electronically. Any transaction approved by the work station gives the merchant a guaranteed payment. When the data has been captured, the job card content is erased. A new and updated hot list is then entered into the job Smart Card to be used for later merchant transactions.

The customer may, at some point in time, wish to see his captured transactions. Perhaps he wants to compare the transaction journal record with his billing statement. He has a user Smart Card reader attached to his TV/Videotex unit at home. Upon insertion of his user Smart Card FTC, a menu appears, offering various inquiry functions. This is the same unit that allows use of the FTC for communications-based shopping, financial transactions, and banking through remote communications.

APPLICATION AND ECONOMIC JUSTIFICATION OF SMART CARD SYSTEMS

In Smart Card systems, there is no paper check to write, process, or return. The merchant checkout process is thus speeded. The merchant need not research a printed hot card list. The bank avoids printing and distributing the paper hot list. Every transaction is authorized against available funds. Electronic capture replaces the labor-and paper-intensive data capture process at the bank. Cash handling and reconciliation activities, cash security, and bank-related activities are reduced or eliminated. Cash register use, card imprinter use, and imprinting form sets (expensive, multi-part, and carbon), are eliminated. There is reduced customer and merchant clerk training needs. Carrying and reconciling paper to the bank is stopped. These are but a few of the advantages resulting from the Smart Card implementation.

All in all, the Smart Card FTC seems exciting! It offers significant new electronic function and control. It seems to have economic justification factors. What is the catch? Today's Smart Cards are expensive compared to the conventional magnetic-stripe FTCs. Personalized Smart Card costs can be as much as twice those of the current bank FTCs. Not all of the answers to the Smart Card operational, security, and control functions have been demonstrated to the satisfaction of the U.S. banking industry. Long-term (3 years) durability of the user Smart Card is now being demonstrated outside of laboratories. There is a strong feeling that the current French banking test merchant card hot list capacity (800 account numbers) is grossly inadequate for the United States environment. Also, as with other new technologies, there will be security threats not yet understood. How will these unknowns be protected against?

In short, we are at an early point in the development of the Smart Card. But issues such as these were successfully tackled before with the magnetic-stripe FTC. Today the conventional magnetic-stripe FTC is reaching several limitations, including limited stripe-recording capacity, weak security, and lack of a plan for further evolution. Thus, seeking solutions in the new communications-based environment will reopen the issue of card technology directions. As detailed needs are worked out, the Smart Card will be a candidate for their resolution.

In the chapters to follow, most of these issues are explored in depth. Application and Smart Card characteristics for a variety of industries are described. The Smart Card will grow in function, shrink in cost, and become the key to a massive usage of information processing by our society.

3 What Is a Financial Transaction Card?

Cards have been with us for ages. Calling cards, playing cards, identification cards, name cards, and place cards are but a few examples. The broad acceptance and use of cards speaks to their functional usefulness. The material used in cards was determined by acceptable economics and expected life needs. Some cards, like paper calling cards, were used in very protected environments. Other cards were used in a wide variety of environments, for example, plastic identification cards and military dog tags.

The earliest reference to a credit card appeared in the 1880s in James Bellamy's book *Looking Backward*. This was preceded by centuries of paper currencies and checks, preceded by ages of barter for shells, stones, and other representations of value. It was out of this historic evolution that the conventional, hand-held, plastic financial transaction card appeared in large quantities in the 1950s. It was introduced by airlines, banks, and the travel and entertainment industries. Initially it was an embossed card only. That is, it had a set of raised characters, used mechanically to imprint the card issuer's and the customer's account number on a paper-forms set.

In the late 1960s, plastic packages for transistors and the early chips appeared. These were used for circuit packaging and insertion into electronic machines. The first references to electronic memory or storage cards appeared in early 1970s patents. Also, in the late 1960s the FTC was further enhanced by the addition of magnetic stripes on the back of the card, intended to provide an economic method to quickly and accurately enter the card-identifying data in a using device such as an ATM. This need was identified in human factors tests that evaluated entry-error rates versus account number length. The customer entry error rate was so high for the 19-digit identification numbers that an automatic technique was the only approach considered workable.

By the mid 1970s, the magnetic-stripe FTCs gained acceptance on a worldwide basis and were being issued in large quantities. Meanwhile, the electronic storage-and-logic card appeared in the patent literature. The magnetic-stripe plastic card is now widely issued and used on a worldwide basis. At some point in the 1980s the worldwide issue will reach 1,000,000,000 magnetic-stripe FTCs. In addition, many of these cards are reissued on a one- to two-year basis. Its wide acceptance, usage experience, and marketplace success suggests broad familiarity and unless otherwise stated, will be used as the example in this book.

In the process of achieving this massive, worldwide issue and use, the FTC has taken on many interesting properties and roles. Since the Smart Card characteristics will, in one of its major uses, be added to those of the FTC, understanding FTC history will be quite valuable.

WHY FINANCIAL TRANSACTION CARDS?

The primary function of the FTC is convenience as seen by the card user, by the merchant, and by the card issuer. An important additional reason is the improved payment facility and services that are routinely available through card use, such as gaining access to a credit line.

Convenience takes the form of faster payment at the checkout stand, easier data capture without writing, and an easier identification of the card holder. From a payment point-of-view, use of FTC credit allows a once-a-month payment of just one bill. It also offers easy use of credit privileges, allowing customers to spend more money more easily, a very attractive factor to the merchant. A debit card transaction will directly reduce the associated savings or checking account.

These convenience features have lead to a variety of FTCs:

Credit: Used against a pre-established line of credit. That is, a preset amount of funds can be borrowed, with a commitment to repayment and appropriate interest charges.

Debit: Used with an existing checking or savings account. These are deposited funds that may be withdrawn with debit card use.

Travel and entertainment: This is a preset limit of funds that may be used, billed, and for which 100 percent payment is then due.
ATM access: Use as a debit card in ATM use only.
Interchange: Use in a multi-institutional environment in which credit and/or debit accounts may be accessed.
Shared: Another type of interchange arrangement in which several institutions share the use of machines, such as ATMs.
Prepaid: Cards for which a specific value has been purchased in advance of their use, such as in telephone or mass transit use.

The FTCs are offered in many industries, including financial services, retail, airlines, and petroleum. The users of cards vary greatly, depending on the type of card. In banking, for example, the following users are generally found:

Credit: The top one third of all bank customers, based on annual income and past credit performance.
Travel and Entertainment: The top 10 to 20 percent of the population, based on annual income and past credit performance.
Debit card: The top one third of all bank customers, based on annual income and past credit performance.
ATM access: Anyone with a savings or checking account, but strictly limited to on-line authorization, based on the customer's checking and savings account balances.
Interchange: Generally the same limit as the credit and debit cards.
Shared-system cards: Generally the same limit as the ATM access.
Prepaid: Anyone with sufficient funds to purchase the card. The card is used to pay for telephone calls or mass transit use.

Overall, it is fair to conclude that there is broad availability of ATM access cards. However, only the top third of the households with banking accounts (85 percent of the United States households) have access to the credit financial card services with its interchange possibilities.

THE STANDARD CARD

The international and United States FTC standards are essentially the same. They have been generated by the International Organization for Standards (ISO) Technical Committee (designated ISO TC 97/17). For the United States, they have been written by a group of the American National Standards Committee (designated ANSC X3B10). These standards have recently been brought up to date in a planned reissue cycle, and broken into a new modularized format. This enables adding new devices, such as the Smart Card, without having to reissue the entire set. This is the manner in which the physical card standards for Integrated Circuit (Smart) Card will be issued, as will be discussed shortly.

The standard modules (ISO TC 97/17) for identification cards type 1 or FTCs are as follows:

Physical card (ISO document 3885/00)
Recording technique—embossing (ISO 3885/10)
Recording technique—magnetic (ISO 3885/11)
Recording technique—location of embossed characters (ISO 3885 /20)
Location of read-only magnetic tracks (tracks 1 and 2) (ISO 3885/30)
Location of read-write magnetic track (track 3) (ISO 3885/40)
Numbering system and registration procedure for issuer identifiers (ISO 3885/70)
Financial transaction card, track 1 and 2 content (ISO 3885/80)
Banking only track 3 content (issued by ISO TC 68—banking) (ISO 4909)

There are a number of other standards that may also apply. For example, standards for optical character recognition (OCR), for the embossed characters, and for the supermarket scanner code or universal product code (UPC).

These FTC standards are truly international. Only two card-issuing nations seriously departed from the initial magnetic-stripe standards, Japan and France. In each case, they have now migrated to the international standards in at least some major part of their recently issued cards. The new interchange cards of all major card-issuing nations now adhere to the ISO standards listed above.

THE CARD LIFE CYCLE

Cards progress through a number of steps from initial design and manufacturing to their eventual use and destruction. Several of these steps result in significant stresses on the card. These steps therefore need to be anticipated in the design of the card and its handling. The typical life cycle includes the following:

Card fabrication: Multi-layer plastics, heat, and pressure
Stripe and signature panel application: Heat and pressure

Embossing: Physical impact and heat
Placing on a mailer: Physical movement and use of adhesives
Enveloping, addressing and physical mail delivery, sorting and handling
Envelope removal and signature application
Wallet, purse or pocket storage, retrieval, and replacement
Use in work stations
Storage (years for the journal access), physical return, or destruction

There will be more discussion of these stresses later.

FTC PROPERTIES

The official properties of an FTC as defined by the ISO cover the following categories:

Physical card: Dimensions, layout, materials, and special features (e.g., signature panels)
Embossing: Location, content and optical recognition codes
Magnetic stripes: Materials, location, recording, and data content
Account numbering: Structure, assigned codes, registration process, and registration authority

The physical card layout is shown in Figure 3.1. The embossed characters are on the front and the magnetic-stripe recordings are on the reverse side, with some rare national exceptions. For example, Japan has placed its previous, non-ISO-compatible stripe on the front of the card for a transitional period. The card thickness was initially set by the imprinters available to the industry and was carried forward into the magnetic-stripe era as the acceptable dimension for automatic teller machine card readers.

The magnetic stripe layout for FTCs is shown in Figure 3.2. Track 2 is the primary recording for FTC use. The PAN or primary account number is the definitive identifying number for both the card issuer and the card holder. The PAN for interchange of cards consists of the following:

1 digit	Industry code for the issuer
5 digits	Issuer identification
12 digits	Customer identification number
1 digit	Check digit for the customer account number
19 digits	Total

MAGTEK

MAGNETIC STRIPE CARD STANDARDS

Credit Cards*

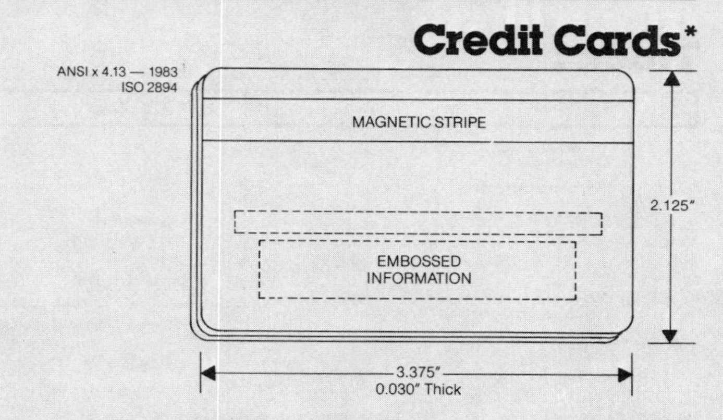

Magnetic Stripe Encoding*

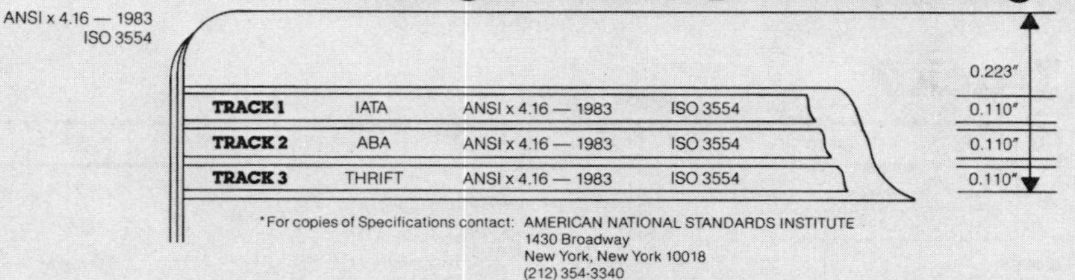

*For copies of Specifications contact: AMERICAN NATIONAL STANDARDS INSTITUTE
1430 Broadway
New York, New York 10018
(212) 354-3340

The Magstripe is recorded with information for automation of transactions

Track 1, developed by the International Air Transportation Association (IATA), contains alphanumeric information for automation of airline ticketing or other transactions where a reservation data base is accessed.

Track 2, developed by the American Bankers Association (ABA), contains numeric information for automation of financial transactions. This track of information is also used by most systems that require an identification number and a minimum of other control information.

Track 3, developed by the Thrift Industry, contains information, some of which is intended to be updated (re-recorded) with each transaction (e.g., Cash Dispensers that operate "off-line").

Figure 3.1. The financial transaction card. (*Courtesy of Mag-Tek.*)

Card Data Format

	Recording Density (bits per inch)	Character Configuration (including Parity Bit)	Information Content
TRACK 1	210 bpi	7 bits per character	79 alphanumeric characters
TRACK 2	75 bpi	5 bits per character	40 numeric characters
TRACK 3	210 bpi	5 bits per character	107 numeric characters

Track 1

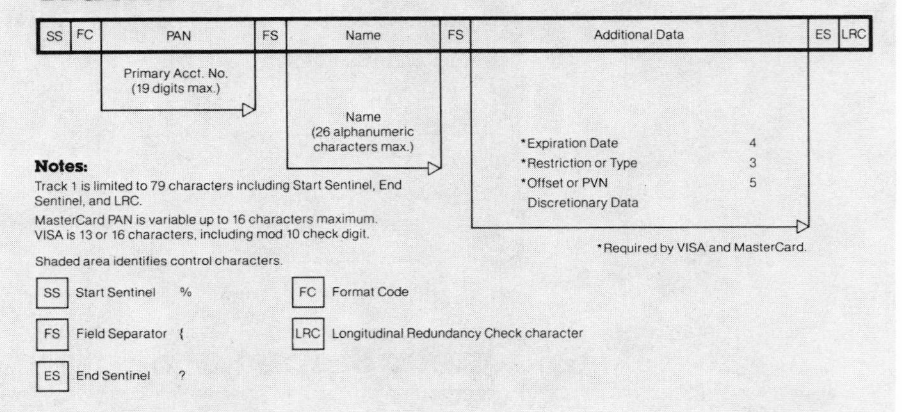

Track 2

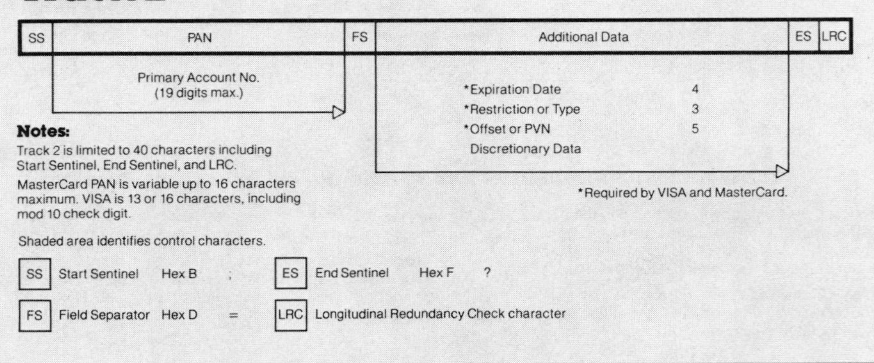

Figure 3.2. Magnetic stripe characteristics of the FTC. (*Courtesy of Mag-Tek.*)

The standard identification numbering system for allocation of numbers worldwide is under the supervision of a registration authority, currently the American Bankers Association. They allocate the 5-digit issuer codes within the 1-digit industry codes. For example, industry codes 4 and 5 are assigned to the banking industry. The overall control and ground rules for the registration authority is set by the international card technology group—ISO TC 97/17.

THE READABLE FTC

The hand-held, user-carried FTC is a multiple characteristic medium. It carries a number of readable characteristics. Some characteristics are human readable, such as: issuer, service, and customer names; and identifying numbers, usage dates, and card manufacturing lots.

Other characteristics are machine readable, such as: magnetic-stripe content, optical character recognition codes, and bar codes.

PASSIVE DATA

These standards cover a broad array of specifications. Most are too detailed for this discussion, however, it is important to note that all of the conventional FTC content is "passive." The information is openly available to all users and handlers of the card. The information remains the same throughout the life of the card, except for track 3. Track 3 is intended for use in applications where the stripe content is changed and is rewritten after most uses. The card offers the data to anyone, including the using person or device. The card does not in any manner interact with or modify the action of the user, unless as a result of the using device's action based upon the card information content. Within the using device (e.g., an ATM point-of-sale device, or a work station), the data from the card is used, processed, and decisions are made. This is generally done in conjunction with data and conditions provided from other sources. For example, card transaction authorizations based on information from a centralized activity data base, accessed by a communications function.

TRACK 2 CONTENT

This conversion of a passive role to a dynamic role for card-based information will be a major change in migrating to the Smart Card. The

Smart Card will introduce its own logic and dynamic interaction with the using device. Therefore, to set the base for comparison, let's first examine the content and purpose of magnetic track 2 as a passive medium. This is the primary track for most of today's direct use of FTC machine-readable information. Track 2 is usually referred to as the *banking industry track*.

The track 2 content and purpose are as follows:

Primary account number (19 digits maximum): Identifies the issuer industry, the issuer, and the user identification number, with the latter having a one-digit check value.

Expiration date (4 digits in year/month format): The date after which the card will not be accepted unless approved by an authorization transaction.

Offset or PIN parameter (Optional 5 digits maximum): This data allows a local PIN value check without issuer data base access when the PIN value is checkable against other information such as the customer primary account number.

Optional track 2 content:

Card restrictions or type (VISA and Interbank, 3 digits maximum): These values limit the use of the card without issuer data base access or independent authorization.

Discretionary data (up to a total track capacity of 40 digits, including 3 prescribed control symbols): Other data may include date of first use or cross reference numbers.

As you may observe, the prime purpose of the stripe content is a quick and accurate entry of the customer's identifying number. If the stripe content provides added data for decision making, then it becomes a more tempting fraud target. As long as the banking industry does not overload the stripe, it will remain under control. However, the industry may need more data. When, why, and how is the subject for subsequent chapters.

WHY ARE FTCs BOTH EMBOSSED AND STRIPED?

Historically, the embossed FTC was well established by the end of the 1960s. Then, the magnetic-stripe card appeared on the scene. The industry had made a major investment in embossed-characters imprinting devices. It would take a number of years before an adequate population of magnetic-stripe readers became available and were put into use. Hence, providing both the embossing and stripe features was a transition technique. It allowed issued cards to be used in embossing devices while the magnetic-stripe devices built up their numbers.

THE FIRST DEPARTURE FROM EMBOSSING

The first example of a nonembossed FTC appeared in the United States. The VISA electron card (Figure 3.3) uses a nonembossed account number, which may lead to some exciting new card trends, such as:

Eliminate paper for data capture: No embossing means no imprinting of a paper-forms set for data capture and receipting.

Require 100 percent electronic authorization and data capture: No embossing requires an alternative technique to enter data and provide a connection to a work station, which communicates with an authorization center.

Enhanced control extends market usage: The 100 percent authorization allows extending card use and credit to a larger number of customers. These are customers who might not have qualified under the reduced control of pre-electron embossed cards. These customers were controlled previously by much less effective techniques such as printed hot card lists.

Eliminated paper processing offers better economic incentive to displace checks and currency at merchants.

It has taken over 30 years to start phasing out embossing and it is not clear when it will be completed, if ever. Hence the Smart Card will start its FTC life as a carrier of both embossed and striped media.

CARD MATERIALS

The plastic material used for card manufacture have some interesting properties. A card is a plastic laminate. The outer layers are transparent and protect the inner core and its surface printing and graphics. The inner core is the prime body of the card. The surface allows for a heat/chemical bonding of other materials to the card surface. For example, the signature panel and the magnetic stripe material are heat bonded into the surface. This assures good adhesion. It also assists visual inspection of the card surface, if the merchant will pay attention. For

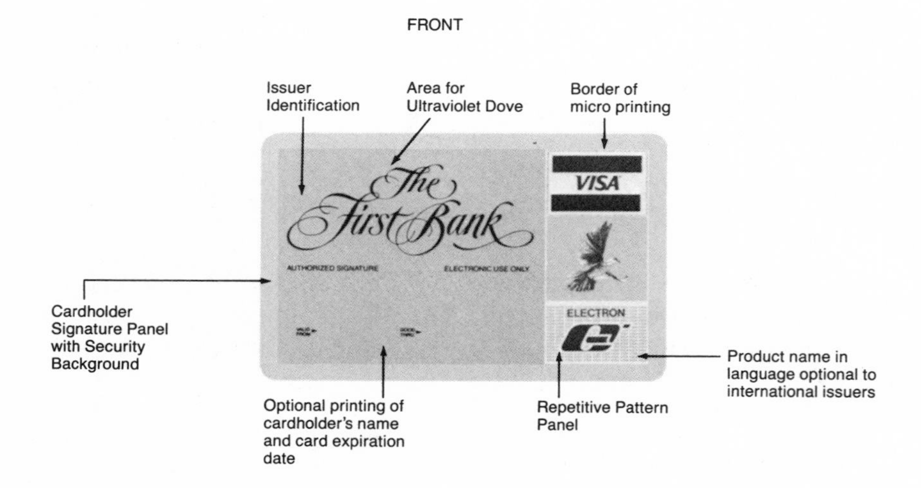

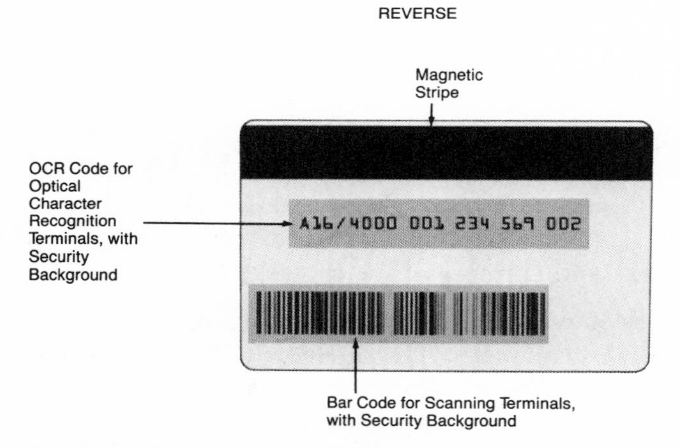

Figure 3.3 The multiple code VISA proposed card. *(Courtesy of VISA International, San Francisco, CA.)*

example, visual inspection can detect attempts to put a new signature surface over the existing surface, or wrapping a length of magnetic tape around a card to fool a point-of-sale unit. Sometimes a decal is added to the surface with a phone number to call for a lost or stolen cards. But of course the decal belongs elsewhere, since it needs to be available *after* the card is lost or stolen.

There is one other surface characteristic of note. The card plastic material is chosen to soften at about 50 degrees Celsius. That is about 120 degrees Fahrenheit, which is a very low temperature. This allows the card embosser to create the raised characters. Unfortunately, the card material is a thermal plastic—the effect can work both ways. Any exposure of the card surface to heat will allow "debossing" of the characters to occur. Cards in a wallet or on a car dashboard on a hot summer day in a hot climate area will be heated sufficiently to deboss, especially if there is surface pressure on the card.

Alteration techniques may use heat to fraudulently deboss and reimboss cards to change the customer identification number. This is the reason for the introduction of holographs on new card issue. They are physically placed over several embossed digits such that if the embossed digits are changed with heat, then the heat also destroys the three-dimensional visual effect of the holograph. That is, of course, if the merchant is paying attention and is trained to discern the visual change. The holograph has improved the ability of a visual detection of embossed character changes, if merchants will take advantage of it.

OTHER FTC PROPERTIES

In addition to standard properties, FTCs have taken on other properties, which are important sources of other concerns that need to be understood. This will be equally true when Smart Card functions are added to those of the FTCs.

Unofficial and Unhelpful Other Roles

Financial transaction cards are usually carried in wallets or purses. As such, they are frequently reached for when other, seemingly unrelated needs arise, giving the cards some interesting roles. Because these other roles usually physically abuse the cards, they may interfere with normal use of the Smart Card. Unfortunately, customer abuse may turn into customer dissatisfaction if not properly anticipated.

Several common examples of this sort of card use include the following:

A car window scraper
A shoe horn
A door latch opener

A toothpick
A luggage tag
A hotel or travel identification device
A package opener
A straight edge
A spacer
A wedge in a malfunctioning imprinter
A dashboard temperature indicator
A perspiration or water squeegee

In severely cold climates, the "window scraper" will be damaged by ice. In Norway, the problem became so severe that the FTC is accompanied by a same size and shape window scraper. I assume that the customer is cooperative.

If an action involves any conceivable form of physical abuse, it has probably been done to an FTC. With the advent of the Smart Card, the abuse potential will be increased. The Smart Card opens a new area of vulnerability with the insertion of an electronic device into the center of the card. Designers must now protect against the problems of internal card heating and heat dissipation, electrostatic discharge, internal and surface conductor damage from bending and torsion forces (I wonder if they'll make a better scraper device?), and amateur experimentation on the electrical contacts and circuits.

Enough! You get the idea. Will it be a challenge for the card to stay alive in this new world? Never fear! Experience indicates that customers will draw a line. They like the convenience and use of the cards. They will not generally cross a line of physical abuse that prevents their use for their intended purpose. As a rule card holders put card utilization ahead of casual card destruction.

The Human Factors Card

The FTC is 2.125 inches high by 3.370 inches wide by 0.030 inches thick (53.98 mm × 85.60 mm × 0.76 mm). This standard size is quite useful for several human factor reasons:

Easily held, oriented, inserted, and retrieved physical device.
Convenient to carry in a wallet or purse.
Offers a wide range of readable graphics and business symbols.
Users can be easily instructed on proper orientation and direction for use in a range of devices.

Can easily carry visual identification of the card issuer, card holder, and card security features (e.g., signature panel).
Effective communication of authorization source, level, and identification numbers.
Human readable text for legal, marketing, and usage needs.

The FTC is not large. However, its "geography" has been allocated by experience. It is truly an internationally acceptable and usable device. Individual users have no difficulty with cards anywhere. However, they do have a language barrier. In some cases, country or language preferences can be indicated on the card or in the stripe information. Smart cards will further add to that facility. (Those new functions will be discussed further.)

Proper human factors consideration in the card design and use is second only to the reliable functioning of the card or its supporting machinery. Namely, that card users will react very negatively to a card or its using device that does not produce its intended results FOR ANY REASON. They'll complain, attack, and abuse inoperative units and will quickly cease being cooperative users no matter how well the human factor design has been done.

In addition to the human factors design of the card, there needs to be a similar concern for the design of the using device: proper lighting and location, well-chosen terminology and physical signs, simple and unambiguous operational instructions. All of these are important human factor elements. When these factors are reasonably chosen and tested they create an easy and inviting user environment. In addition, the method of customer introduction to the device operation has some critically important considerations. These will be covered later in Chapter 9.

The Legal Card

Any medium that has a payment role requires some consideration of legal matters. A legal text appears on the reverse bottom portion of FTCs. However, there are many other legal functions.

Legal identification of the card issuer, card holder, and issuer industry
Trademarks and copyrights as used on the card
Issue and acceptance of the card conditions
Conditions and point-of-return for cards
Period of use, including start and end dates
Identification of the card owner (usually the issuer)

Notice of need to return the card on demand
Country of card manufacture and origin
Nontransferability of the card
Invalidation of card use upon notification
Signature of user as an indication of acceptance of the terms and conditions
Notification that any use of the card is acceptance of the terms and conditions
Agreement in advance to future amendments to these agreements
Action required if the card is lost or stolen
User's potential liabilities for authorized or other card use
Exclusion of liabilities after proper notification to issuer
Currency of settlement for card transactions
Country laws designated for legal adjudication

Some countries, such as Sweden, have gone so far as to treat FTCs in a legal manner equivalent to the national currency. This is especially applicable to issues such as card counterfeiting. Interestingly, most countries have not issued rules prohibiting card alteration, counterfeiting, or other plastic abuse. Perhaps this will change in the future.

Card Fraud

During the first decade of magnetic-stripe FTCs issued in the United States, the losses for VISA and Master Card have averaged about one percent of their gross sales. In 1986, sales were just over $150 billion, so losses were close to $2 billion. Generally, over the last decade, 85 percent of the losses were to bad debt. That is, the person to whom the card was issued did not pay the bill. The remaining 15 percent of the losses was due to card-related crime.

Recently, these losses were divided as shown in Table 3.1

The sudden increase in fraud has caused a focus on the anticrime features of plastic cards. The recently announced VISA proposed card best portrays the latest set of FTC features to combat these losses (Figure 3.4).

Holograph
Fine background print
Nonembossed identification numbers
Signature panel on the card face
Signature panel that shows VOID when altered
Ultraviolet light energized dove outline under the issuer's logo

TABLE 3.1

	LOSSES		
	PERCENTAGE	VALUE (in millions of dollars)	
Merchants	22.6	37.7	(e.g., collusion)
Stolen	17.2	28.7	
Counterfeit	14.3	23.9	(up 430% since 1980)
Altered Embossing	8.1	13.5	(up 86% since 1980)
Lost	7.5	12.5	
Phony Applications	7.4	12.4	
Telephone Orders	6.3	10.5	(e.g., wrong account number)
Mail orders	4.5	7.5	
Postal theft	3.6	6.0	
Manufacturing and Mailing	2.9	4.8	(e.g., physical theft)
Cardholder collusion	2.1	3.5	(e.g., denial of use)
Other	3.5	6.0	

Anticopy background on retail (OCR) and supermarket (UPC) codes
Other features which have not been described

In addition, FTCs offer other security and identification features, including:

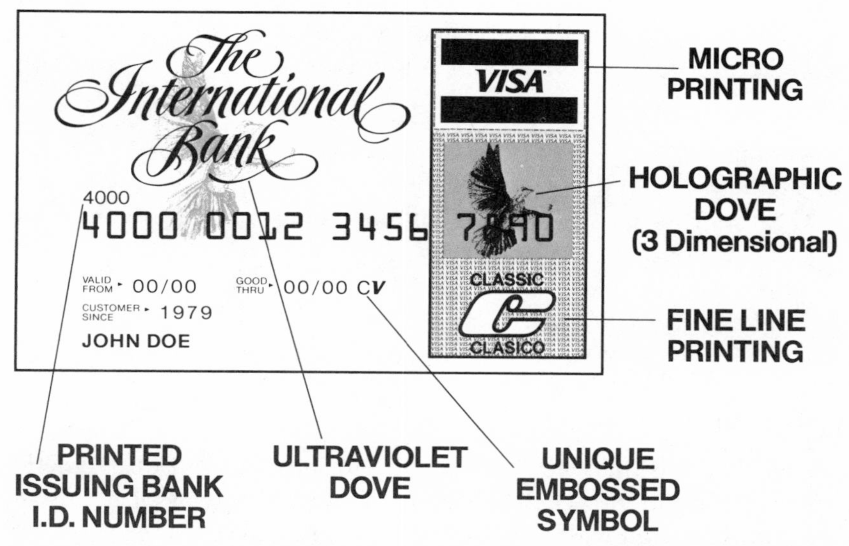

Figure 3.4. The VISA proposed card anti-crime features. *(Courtesy of VISA International, San Francisco, CA.)*

Personal identification
  Signature
  Name
  Photograph (sometimes)
  Fingerprint (sometimes)
  Employee Number (sometimes)
Date of use expiration: Stops usage on a lost or stolen card
Issuing organization: Identifies the authorization point
Color logos, fine printing background: Anticounterfeiting devices
Card width, shape, and rigidness: Operates antivandal gates on ATM card readers.

There are other announced features in the marketplace for anticrime features of plastic cards and magnetic stripes. Most features have significant weaknesses when examined by experts. It is not our intention to provide a crime primer. Suffice to observe that (1) there are no secrets; (2) there are very inventive opponents out there; and (3) any good security design includes sensitivity tests to detect unknown attacks, e.g., card usage frequency and dollar use limit checks.

Magnetic-Stripe Exposures

In addition to the plastic features, there are the magnetic stripes. No claim has ever been made for magnetic-stripe security. In fact, most knowledgeable tape experts readily admit that the magnetic stripe content is:

Readable
Alterable
Modifiable
Replaceable
Refreshable (original content can be replaced, e.g., track 3 data base)
Skimmable (can be quickly and illegally read)
Counterfeitable
Erasable
Simulatable

Why then has there not been a more serious incident of magnetic-stripe-based fraud? Magnetic-stripe reading is usually associated with direct authorization with positive files. That is, card usage requests are checked with centralized account records and actual overall card activity before most significant transactions. In numerous transactions such as unattended ATMs, a personal identification number (PIN), is used to augment the magnetic-stripe data in making an authorization decision. The inclusion of the issuer identification in the magnetic-stripe data assists the card transaction "acquirer" to know which authorization center to call for control of the card transaction. This control concept has, so far, detected most card frauds when the transactions are large enough to warrant on-line control.

How would the industry limit a runaway counterfeit fraud problem? Most authorization systems check for frequency of card use in a day. For example, more than three uses in a day will stop authorization until the issuer is satisfied that the card is in the possession of its legitimate holder. Even so, most issuers will also limit the total dollar usage of a card in one day. These sensitivity tests help to limit losses from counterfeit and simulated magnetic stripes. They work because the control point is positioned to see the combined action on a card. This will be a major security challenge to the Smart Card.

Overall, the anticrime efforts have been a modest success. The sensitivity checks have really been the important features, but even they need a few "hits" before an alarm is sounded. Smart Cards have the potential for better security, to be discussed in a later chapter.

The Interchange Card

Great care and expense has gone into the graphics design of major interchange cards like VISA and Master Card. They each have unique geometric designs and carefully selected colors and placement of names and symbols. All of these factors are designed to provide quick recognition, to achieve merchant association, and to encourage card usage.

Only recently has the overwhelming use of the full card front for the interchange logo given way to space for the issuer's identification logo. In fact, the recent announcement of the VISA proposed card (see Fig. 3.4) provided for the issuer's name and logo to reassert itself as the most promiment identification on the front of the card. This permits individual issuers or market interchange arrangements to assert their own identification as part of an effective marketing program.

To demonstrate the importance of the graphics, let me recount an early incident in a standards decision on the magnetic stripe on bank cards. In the late 1960s these standards were being finalized. An important, but open, question was whether the magnetic stripe should be on the front or rear of the card. Several samples were made for each layout. The chairman of the industry standards group was the chairman of a major West Coast bank. His bank had just spent millions to advertise their interchange card graphic. Placing the magnetic stripe on the front of the card would result in a one-half-inch blacktop road across the newly promoted logo. The only alternative would be to reduce the size of the graphic. The committee chairman saw the front sample. He stated simply that he was not about to reduce the size of his "circles." The decision was made. It has never been reopened. The stripe is on the back of the FTC. The importance of the front graphic continues to be respected by the industry.

There are a number of important graphic and symbol considerations in the card design:

- Graphic trade marks and copyrights
- Manufacturing batch numbers, control codes, and control dates
- Signature panel for user identification verification
- User photograph
- Printed text to identify the embossed characters
- Interchange arrangement graphics
  - Shared ATMs
  - Shared point-of-sale
  - Interstate and intrastate banking facilities
  - Card interchange
  - Other interchange services
- Card manufacturer
- Other codes and graphics (OCR and bar codes)

As shared and interchange arrangements continue to grow, the symbols on a card will also increase. I know of one case where an issuing bank has six symbols on the back of the card. They expect the number of symbols to reach sixteen. This results in a very crowded surface—usually on the back of the card.

SUMMARY

Thirty years and one billion FTCs later! This is the inherited legacy of the Smart Card. However, as it must to all technological achievement, progress means new capabilities. Those new capabilities of the Smart Card will be added to the conventional FTC.

4 The Physical Smart Card

The Smart Card is a combination of a package, one or more integrated circuit chips, and a using or interface device.

In the case of an FTC, a Smart Card is an FTC into which one or more integrated circuit chips have been placed. The FTC maintains all its original characteristics, with some additions.

There is an electrical connection to the chips, achieved by contacts or a contactless equivalent, providing for both power and signals. Power for the chips usually comes from the interfacing devices, with the correct voltages, frequencies, and current characteristics. The signal interface provides both data and logic information flow with correct voltage levels and frequencies. The using or interface device interacts electronically with the Smart Card logic and storage content to achieve a job or application result.

In this section I will discuss the Smart Card itself. In subsequent chapters I will discuss its application, system, and interaction characteristics, including the additional elements needed to use and to fully achieve the Smart Card potentials.

THE PACKAGE

Many considerations will influence the Smart Card package design. The most significant packaging factor will relate to the need to support existing devices.

An Added Function

In the case of the financial transaction card, the Smart Card functions are being added to an existing medium. For the card to be usable in a large set of existing using machines, it is preferable to keep the same package. A new card must be equally usable in present generation ATMs as well as in next generation, stand-alone merchants' work stations.

The thinness and flexibility of the Smart Card will be a real technical challenge. Chips are produced on a relatively thick silicon base. The base is also very rigid and may be brittle. Conductive leads will connect the chip circuits to the surface contacts. This attachment to the chip takes vertical space. The leads also need to be flexible enough to survive in the bending card. Cards are subject to torsion or twisting forces. In addition, the card is subject to impact forces when the card is embossed or used in an imprinter.

A New Device in an Old Environment

The use of a Smart Card in the military will create a new package called a *data tag* (Figure 4.1). It will be worn around the neck for ease of carrying, retrieval, and access. However, it will be thicker than the conventional dog tag, to simplify the package design tasks. The data tag will need a set of electrical contacts and must be able to resist dirt and be easily cleansed. Also, since it will be worn around the neck, it must be sufficiently rugged to survive a wide range of battlefield abuse, including hot showers and cold water immersion.

An Entirely New Card in a New Environment

Consider, for example, the use of Smart Cards in long distance telephone calls or mass transportation. The Smart Card might be used to hold sufficient value usable to make multiple purchases. These cards might be rigid, narrow (1–2 inches), and not too thin. They would be elongated (3–5 inches) for insertion into an interfacing device. This configuration would need to combine an ease of pocket carrying and, when vending unit dispensed, to maintain its usage reliability and ease of operation.

A Smart Card for a television descrambling device might be no larger than a matchbook cover, enough only to provide a surface for the contacts. The Smart Card size would be set to enable the user to insert and remove the card from the descrambler for the valid period, e.g., a month.

Figure 4.1. The Military data tag Smart Card. (*Courtesy of Datakey Corporation, Burnsville, MN.*)

A medical profile Smart Card would need a very large storage capacity. It would also require a self-contained power supply in order to use a very dense type of storage technology. The power and number of chips might require a Smart Card 3 to 10 times the thickness of a financial transaction card. It would also need to be easily seen and accessible in the event of an emergency. It might require a very reliable set of contacts with a protective cover, necessary if an on-line source of data was not available at the scene of an accident.

PACKAGE PARAMETERS

The challenge in the package design will be in containing and protecting the chips, and in offering reliable electrical connections. The contacts will need to be added in a manner that:

Achieves the least abuse and contamination of the contacts
"Watches" the side of the Smart Card package opposite the contact surface to avoid surface deformities
Avoids disturbing or distorting the package dimensions or flatness
Maintains the expected package useful life and wear characteristics

The shape and thinness will be set by a number of factors:

Human factors of handling, insertion, and removal
Compatibility with prior cards and card-handling devices
Impact of package dimensions on chip(s) life
Need to coexist with widely used industry standards
The desired electrical properties, e.g., a sufficient thickness for a battery supported technology
Sufficient geography and surface to handle the marketing, legal, and human factors legends

An added packaging consideration is the material used. Adding one or more chips to a conventional FTC means using the same materials. However, the plastics involved may release certain contaminating vapors or plasticizers, which need to be protected against.

CHIPS IN THE PACKAGE

There may be some important material factors or contaminants to be considered with chip insertion. The chips will probably need to be hermetically sealed. Sealing is necessary to allow the chips to operate in the presence of water and many types of contaminating vapors, chemicals, and solvents. For example, self-service units in gas stations are subject to water and chemical fumes, which may damage the connectors or the leads or the chips or all of them.

HEAT

As an electrical device, the chip or chips will need to be able to dissipate heat it generates when the using device power and signals are applied.

The amount of power and its duration will cover a broad range of heat generation. The heat is also a function of the chip technology, capacity, and use. For example, a chip with dense content will experience heavy heat buildup during its program loading and personalizing process.

Heat dissipation maximum allowance and associated temperature levels will be specified in the standard for the card. For example, the card surface temperature should not be allowed to exceed 50 degrees Celsius (120 degrees Fahrenheit). Remember, that was the temperature at which the card surface material starts to soften. Any higher temperature will distort the card. In fact, if the surface becomes too warm, the chips might pop out of the surface. This limitation may require a heat dissipation element in the using device, e.g., a radiator surface around the chip.

ELECTROSTATICS

Another concern in the Smart Card package is electrostatic charges. If cards are used in certain environments, there may be a large static electricity buildup. The charge buildup may distort or destroy normal chip operation. In fact, the action may be sporadic and inconsistent. A dry room with thick rugs in a warm environment might be very sufficient to generate damaging levels of electrostatic charge. In fact, some home video games in a similar environment have had a problem of this type. Hence, static conductive coatings might be needed to minimize this exposure, together with an appropriate work station grounding mechanism. Testing specifications for this issue are contained in the card standards (ISO TC 97/17/4).

THE CONTACTS

The contacts provide the electrical connections between the internal chip(s) and the external using device. The current draft standard for the Financial Transaction Smart Card (ISO 97/17/4), provides for 8 contacts on the front or back of the card. The choice of the back or front contacts for the Smart Card is at the issuer's option. These contacts each have a face of 2.0 mm × 1.7 mm (or 0.079 in. × 0.067 in.) and are flush with the plastic card surface. They have a maximum displacement between the lowest and highest contact of 0.1 mm (0.004 in.)

It is already known that certain contaminants like dried glue can block electrical conduction. Smart Cards will be placed on mailers for delivery to the user, and if care is not taken, "dried glue" contact contamination may occur. To avoid such contamination, several contactless Smart Cards are being demonstrated. One type uses inductive coils. Another uses small, high-frequency transmitting and receiving antennas to move power and signals to and from the Smart Card. Carrying financial transaction cards also has created problems. Accumulated dirt and contaminants from poorly applied signature panels will adhere to the contacts. Contact probes on the card interfacing devices must be chosen to operate through contaminants, contacts such as pin or wiping probes.

The initial industry standard will provide for the contact version. When a reasonably economic and reliable contactless version appears, then a standards module will be provided to accomodate it. Until then, keep your contacts clean, uncontaminated, and avoid contact exposure to chemical eroding agents.

ENVIRONMENTS

There are a number of interesting environmental parameters that need consideration. One is the temperature range at which the chips do not operate. Some cards are stored at external temperatures, inside an automobile in the winter. Another concern is the environmental gases and liquids to which the card is exposed. Since the standards will probably prescribe hermetic sealing of the chip, this may not be an issue. However, sealing does not assure that the outside of the chip insertion mechanism, e.g., an enclosed container, will not be adversely affected by contaminants or temperatures.

The interface to the chip is a set of eight front or back contacts. These contacts apply a range of voltages to the card. In some environments the interfacing device may not be operating correctly. It may leave all voltages applied to a set of wiping contacts. As the card is inserted or removed, higher voltages may be "wiped" across the wrong contacts. Card designers need to anticipate this possibility. It may be countered by placing the high voltages on the contacts closest to the insertion card edge. Thus, as the card is inserted, the high voltage contacts are the last to touch the card surface contacts. Another alternative is to provide a work station interface interlock circuit that turns power off until the Smart Card is fully inserted.

PHYSICAL FORCES

The method of attachment of the chip to the external contacts is important for several reasons. There is not much "face space" above the chip and below the card surface. Thus, an attachment technique that lowers this profile is necessary. The attachment also requires a method that will survive the various twisting and bending forces on the card. As the number of chips is increased, or the distance from the chips to the contacts becomes longer, the forces at work become more significant.

Physical forces working on the Smart Card can be quite sharp and severe. A serious example is the impact force of an embosser. The embossed line of an FTC comes within 3mm (0.118 in.) of the contact area—and so the chip location. The embosser mechanism strikes the plastic card with sufficient force to cause the plastic to heat and flow into the embossed character shape. It strikes the card for each character. That might be up to three lines of twenty characters each, or a maximum of sixty times. This creates a manufacturing dilemma.

Conventional FTCs are assembled in large sheets for manufacturing. The sheets facilitate a multi-layer, heat-and pressure-based lamination process. The individual cards are cut out of the sheets then completed with the application of signature panels and magnetic stripes and personalized with embossing and magnetic-stripe encoding.

If the embossing process creates unacceptable forces on the chip(s), then the manufacturing sequence may need to be modified. For example, several vendors are suggesting that the chip be placed in its own self-contained package, including the contacts. This would provide an hermetical sealing, flat and short connectors, mechanical protection, and heat and electrostatic shielding. The physical card would be designed for sheet production. When the cards are cut out of the sheet, an insertion area could be prepared for the chip module insertion. After this, the card will be personalized with embossing and magnetic stripe recording. Then, the chip module could be inserted and personalized.

This also avoids the embossing forces on the chip. This might avoid the issue of dealing with chip cards that have embossing rejection after chip insertion. Throwing away chips could be expensive—at least in the earlier, expensive-chip days. It is not clear that an economical manufacturing plan has yet been demonstrated. However, necessity is the mother of invention and some reasonable solution will emerge from future industry experience with this medium. Not embossing the financial transaction Smart Card may be an acceptable solution. This solution will depend on the acceptance of alternatives such as the new VISA non-embossed proposed card.

THE INTEGRATED CIRCUIT CHIPS

The information processing properties of the chip is the major new attribute of the Smart Cards. There are a broad range of applicable chip technologies. We'll not discuss the technical alternatives, which are subject to rapid change and improvement; rather, we will consider the parameter characteristics that will influence the technology choices.

Performance: Speed, complexity, capacity, and density versus size.
Power: Levels, circuit needs, capacity, heat dissipation, and self-contained supply.
Design options: Function, security, availability, on-card versus off-card function, and general purpose versus special purpose design.
Trade-offs: Capacity versus cost, performance versus cost, capacity versus useful life, single- versus multiple-chip solutions, and chip thinness versus card reliability.

The range of technology choice is very large. The real challenge will be to understand the Smart Card application and its economically justifiable function. Multiple chips provide larger capacity and the possibility of using general purpose chips. However, larger or multiple chips increase vulnerability to bending or twisting forces. (A chip economic assessment is discussed in the *Economics* chapter, page 136.)

CURRENT SMART CARD TECHNOLOGIES

There are several approaches used in the current French Smart Cards. An 8-bit NMOS type microprocessor is used in Cii Honeywell Bull and Philips Smart Cards. Flonic Schlumberger and Cii Honeywell Bull use tailored chip designs. Flonic Schlumberger has 4.6K bits of electronically programmable read-only memory (EPROM). Cii Honeywell Bull incorporates 8K to 12K bits of EPROM. The Philips card used two chips. These are a microprocessor and a 16K bit EPROM. Ths issuer has 12K of the 16K bits available for application and user memory.

The EPROM is a nonvolatile memory. That is, it retains its stored value while the card is out of its interface device and away from a power supply. Future units from other vendors are expected to include electronically erasable programmable read-only memory (EEPROM or

E^2PROM). These units are electrically alterable as compared to the ultraviolet erasure of EPROM. Also, the memory capacities of future units are expected to increase to 64K bits.

The EEPROM offer a functional ability to rewrite selected data fields. This function must be carefully used so as not to accidentally or deliberately destroy vital data. This renewable application facility has a number of important uses. Also, the added Smart Card capacity may be used to reduce the work station costs and complexity. For example, the Smart Card logic may produce a communications message directly in the attached system formats and prototocols.

CHIP PRODUCT FUNCTION

The range of product function will be dictated by the Smart Card application and, limited by the economic and physical properties of available chip technology. Some Smart Cards will be simple storage devices. They will use the storage units in their operation. Others will require both storage and logic. The FTC will use chip product function to interact with a range of interfacing devices. These will vary from simple phones to complete intelligent and processor driven work stations. Some Smart Cards will be in self-contained work stations. That is, a work station will have one or more tailored Smart Card chips integrated into its physical and logical work station package. This combination will be particularly valuable in portable or "knee-top" intelligent work stations.

The variations in chip properties to be considered include:

General purpose versus special purpose or tailored logic
Fixed versus variable instructions sequences
Instruction capacity
Data representation (e.g., 4-, 6-, 7-, 8-, or 10-bit characters)
Security circuits, algorithms, and controls
Input/output interface
Communications interface and options
Single- versus multiple-chip configurations

THE "TAILORED" CHIP

A chip with its logic and storage tailored to a specific application achieves several important advantages over a general purpose chip:

Eliminates unused general purpose elements (hardware, software, procedural, and servicing).

Allows use of special instructions and data types. These may be more powerful than general purpose types.

Combines logic and data storage into one chip design. This optimizes use of the chip geography.

Adds special on-chip features such as chip and vendor identification, diagnostics, special logic, faster architecture, and voice features.

Developers can use new and more powerful development and diagnostic systems including special high-level languages.

Can utilize design options available for faster functional blocks, increased parallelism, more versatile system interfaces, and special purpose functions.

All these advantages provide increased capacity and performance with a given chip size and technology. These are then added to the growing chip density and falling chip costs. Together, these factors suggest an almost endless capacity to further tailor a broad variety of Smart Cards. There are, however, advantages to using general purpose chips. Their costs benefit from larger production volumes. Applications not having large volume use might economically benefit from the large production volumes with a general purpose program in a generic smart card. On the other hand, the tailored chip may benefit from better performance or reduced capacity advantages.

DATA STORAGE MODES

There will be a wide range of storage operating and accessibility modes, and a variety of recording and reading modes. The selection of these options will depend on the Smart Card application and security needs. The recording and reading modes are shown in Table 4.1.

These data storage capabilities provide an important variety of functions, as seen from the external contacts or by the internal logic. These storage functions provide the following Smart Card capabilities:

1. Data readable externally at all times: for example, the identification number of the card.
2. Data readable externally after appropriate logical or security controls are passed: for example, the available balances and use of an FTC after the correct PIN has been provided to the card.

TABLE 4.1

RECORDING	READING	USE
Record once	Multiple read	Personalizing data
Record once; then destroy recording circuit	Multiple read	Nonalterable data
Record once; then destroy recording circuit	Single read; then destroy record content	One-time-use data
Record once; then destroy recording circuit	No external-read circuits	Internal use only data
Rerecordable	Multiple read	Updatable data

3. Data never readable externally, for example, the PIN that the Smart Card will accept as the correct value.
4. Data that is permanently recorded and is readable externally: This might be a sequence of instructions used to control or diagnose the status of an external device, or it may be the educational record of a soldier, or the transaction log for an FTC, or the previous electrocardiograph trace of the medical Smart Card holder.
5. Data that is selectively updatable under internal logic controls: For example, variable or updatable dollar control fields in an FTC. This may be telephone or communications addresses used in Smart Card applications, the PIN that the card will accept, or a comparable control number that allows the issuing institution to read the card content.

CHIP CAPACITIES

A critical dimension of Smart Card design and use is its logic and data storage capacities. For example, an important function in a Smart Card is the recorded transaction journal. This is a permanently recorded journal. Hence, when the storage capacity for the journal has been filled, the Smart Card is no longer usable.

The Smart Card standard for FTCs will probably seek to specify a maximum expected card life of three years. The current French banking test cards offer a journal capacity of up to about 200 entries. Each of these entries is very small—perhaps 40 bits, or 8 digits, of information. A 200-entry journal would then require 40 times 200 or 8,000 bits. In the current French banking test cards, this is a large storage requirement. And 40 bits per journal entry is very small. Consider the following desired journal entry:

5 digits—Date (DDMMY format)
5 digits—Merchant identification
3 digits—Transaction amount (dollars only)

This transaction journal entry requires 13 digits or 65 bits. Is this enough capacity? Assume the minimum required capacity is 7 transactions per month for a 36-month period. This is 252 journal entries; 65 bits times 252 entries requires 16,380 bits.

To these journal storage bits must be added the storage requirements for logic and operating data. If operating data is 500 digits for identification, services, security, and limits, then an added 500 times 5 bits or 2500 bits will be needed. A microprocessor is needed. A logic program of several thousand steps is needed. Altogether, the capacity needed will approach 32,000 bits minimum. (The capacity costs for these combinations will be discussed in the *Economics* chapter, page 136.) Amazingly, these storage requirements appear to be within the capacities of current and projected chip technologies.

INPUT AND OUTPUT FACILITIES

Some Smart Cards will NOT be cards at all! They will be portable work stations with Smart Card functions. Several current hand calculators are the size and thinness of a financial transaction Smart Card. The calculator includes a decimal keyboard, function keys, a solar panel for energy, and a liquid crystal display. In addition, the hand calculator contains the chips that operate the calculator. Consider providing a Smart Card work station with input and output functional capabilities. This might be done by adding a chip with Smart Card functions and the interfacing contacts to this calculator. Perhaps two of these "financial transactors" or Smart Card work stations could be directly connected to each other and perform a funds transfer. This could be very important. For example, this connection between a user's Smart Card financial transactor and a merchant's cash register might eliminate the need to carry cash. In fact, if this card could be updated remotely, such as through a videotex unit, then it might reduce the demand for automatic teller machines. Why travel to an ATM to get cash? The electronic equivalent of cash could be obtained remotely.

Thus the Smart Card with input and output facilities may be almost any combination of portable work station functional units.

THE SMART CARD COMPARED TO THE PERSONAL COMPUTER

The Smart Card capacity is measured in bits. The personal computer capacity is measured in characters. For a given size, the personal computer storage is quantitatively about ten times greater than the Smart Card. Why the large difference? The personal computer is a general purpose computer *system*. It generally needs to provide storage for the following elements at all times:

An operating system to control the full range of input and output devices or functions that might be called upon.

A complete application program including all of the options, features, and functions that might be called upon.

The applicable data base for the full range of application and operating programs.

Any utility programs or devices necessary during the operation, including communications, display, graphics, and related facilities.

The Smart Card is not a general purpose computer system. It is a completely tailored and personalized control device. It carries a specific set of information content and protocols to be used in a specific application such as a point-of-sale financial transaction. The using devices or system and the other job or device Smart Cards also contain logic and data storage, so the user-carried Smart Card does not require the capacity of a personal computer. There will be a multiplicity of Smart Cards. Each will be highly personalized, and optimized to the specific application usage. The Smart Card will be used in a highly formatted protocol environment. The user will become accustomed to a consistent human factors response across a broad set of applications.

Altogether, the net impact of these factors will be the ability to substitute Smart Cards for the capacity of a general purpose personal computer on an application by application basis.

THE GENERIC SMART CARD

Smart Cards will vary by application function and by product function. This combination provides a broad range of Smart Card configurations and opportunities. The opportunities have been observed in almost every industry group. The challenge is how to best describe this range.

What I call an *opportunity matrix* will be used to describe the range of Smart Cards. The first part—product function—has already been described in the chapter and can be portrayed as shown in Figure 4.2.

This matrix will be added to in the next chapter with the range of application functions.

Basically, Smart Card product functional levels can be described as: Level 1, Storage-only device; Level 2, Storage and logic; Level 3, Storage, logic and input (e.g., keyboard), and output (e.g., display) devices.

An example of each Smart Card product functional level is as follows:

Level 1: A prepaid telephone payment card, containing security code, identification number, and telephone call increments not yet used. In this type of using device, the logic and input/output capabilities are provided by the using unit such as a telephone instrument. The telephone would include the logic for checking card security and content acceptability. It would also have an appropriate input/output device to show or voice the available card balance for the card user.

Level 2: A financial transaction Smart Card. This Smart Card has been described previously. It includes logic for security and application purposes. However, the Smart Card requires that the using device (e.g. a point-of-sale unit) provide the required input/output facility. The using device also includes application and security logic as required to interact with the Smart Card.

Level 3: A portable work station with Smart Card capabilities. The work station has all of its normal function and facilities. However, to those facilities are added the information protocols found with the Smart Card product functions; for example, identification of the workstation, a description of the services and limits to which the work station has access, security protocol, and a journalizing of the transactions into which the work station has entered. These work stations may be the rectangular dimensions of an FTC but probably thicker for design and physical capacity reasons.

The technical capacity to achieve a level 3 Smart Card work station with the physical properties of a financial transaction card are

APPLICATION	PRODUCT FUNCTION		
	CARDS		WORK STATION
FUNCTION	STORAGE	STORAGE + LOGIC	STORAGE + LOGIC + INPUT + OUTPUT

Figure 4.2.

demonstrated by the new Casio card "thinness" calculator. This bank card size calculator is shown in Figure 4.3.

These product functional levels will be discussed further in the next chapter. It will expand the description to show the application levels that match the product functions in the "opportunity matrix."

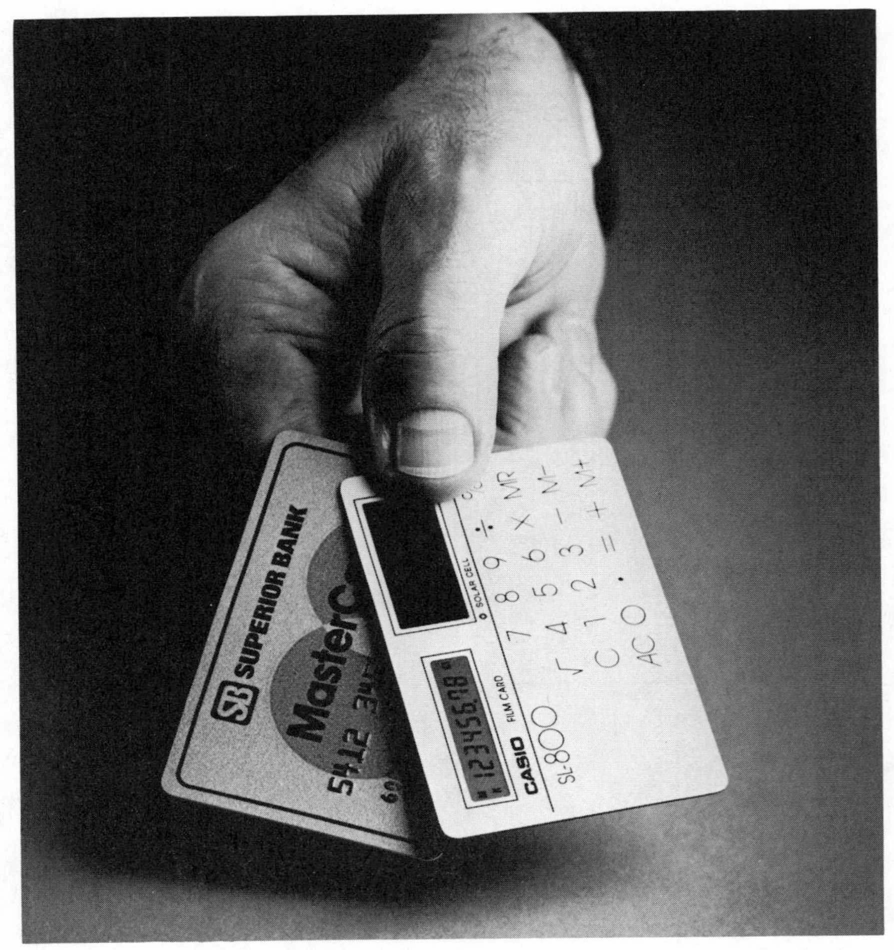

Figure 4.3. The Casio "credit card" calculator. (*Courtesy of Casio, Inc., Consumer Products Division, Fairfield, NJ.*)

SUMMARY

The Smart Card is a physical package. The physical package will be tuned to its application, use, and user. The Smart Card is an information processing device using integrated circuit chip(s). The chip may be general purpose or tailored to the Smart Card application and usage. The Smart Card is a selected set of product functions, which will be some combination of storage, logic, and input/output capabilities. The Smart Card may take the form of a portable medium or it may be a work station with integrated Smart Card functions.

5 Smart Card Applications

Applications are important. They are useful solutions or results, identified by industries or users or functions, achieved by using the application functions that are built into the Smart Card logic and data. To prepare a solution the application functions need to be selected, combined in a development effort, and then applied to demonstrate the expected results.

The objective of this chapter is to describe the range of application solutions available for many industries with Smart Cards. Most of these solutions have not yet been demonstrated with Smart Cards. Rather, this is a projection of possibilities based on current, limited Smart Card tests and extensive experience with microprocessor-based application solutions. The application solutions will be defined by a set of *application functional levels*:

- Write-once data
- Write-once data plus write-once journal
- Write-once data plus write-once journal plus read/write updatable data

These application functional levels will be used to complete the opportunity matrix started in the previous chapter.

The second dimension of application solutions is the information types used with Smart Cards. These information types are found in any and all of the application functional levels. These information types will be discussed in detail to offer a comprehensive guide to Smart Cards application alternatives.

The opportunity matrix will be completed on a generic function basis. Completed matrices for several industries will be shown and discussed to demonstrate Smart Card application repeatability across industry boundaries. This repeatability is vital as it helps to increase the volume of Smart Card opportunity by type of Smart Card. Volume usage produces lower costs per card. Lower costs allow lower prices, which results in better application economic justification. Hence, this structured examination of the opportunity matrix will guide the development of future Smart Card applications.

APPLICATION FUNCTIONAL LEVELS

Application solutions in most industries have many components. Data, logic, system connection, system controls, and others. To describe these application solutions, two dimensions will be used. The first is the application function or level. These are the primary methods of entering and using the Smart Card content. The second dimension is the information type. This is the variety of data and logic that make up the Smart Card content. The application functional levels have the attributes described below.

Write-Once Data

This is data that is entered into prescribed Smart Card content locations. The content may be entered as part of an initializing, personalizing, or adaptive process. This content may be read numerous times. Some of the content will be used internally in the card for logic and application processes, some will be provided to external devices.

An important option with this application function is the ability to destroy the stored data or to destroy the write or read circuits. The first destruction—of stored data—results in the "consumption" of the data. For example, using the data as payment increments in a telephone payment card application. The second form of destruction prevents any further change or modification in the recorded data. In addition, the Smart Card logic design may prevent any external reading of this data.

This is an important security feature. It might protect against external examination of the value of the PIN which the Smart Card will accept as the correct entry.

Write-Once Data plus Write-Once Journal

This application function differs from the previous one, although it also involves write-once capabilities. The journal is a historic record of the card activity. The captured content will have been predefined. The logic will be set to gather the correct information, to format the journal, and to possibly protect its content from false entry and unauthorized modification. The permanent journal record captures a variety of information. Transaction results, external status reports, captured data to update or augment the initial data, and other events—internal and external—add an important application solution dimension. The key attribute of this application function is the ability to make a permanent record of Smart Card usage and involvement.

Write-Once Data plus Write-Once Journal plus Read/Write Updatable Data

This application level adds an interactive application function mode. Updatable data opens the Smart Card application spectrum to jobs which require dynamic response to new data, circumstances, status, events, or logic. This level is achieved without consuming added Smart Card capacity beyond the space allocated for the dynamic interaction area. However, interaction also allows the user to replace data. This mode must be selected carefully based on an assessment of the user's ability to understand and handle this added exposure.

APPLICATION INFORMATION TYPES

The application function levels are used in many industries for a variety of information and logic needed to achieve the Smart Card results. The various information types may appear in each of the application functional levels. The level selected depends upon the application needs.

Identification

This data type consists of three areas. These relate to the Smart Card issuer, the Smart Card user and the Smart Card usage acquirer. The user may be the card holder, an applicable device, or a preset block of Smart Cards. The acquirer may be an institution, an industry, a network, an authorization or control point, an interchange group, or even a country. Identification may be in the form of numbers, names, personal identification number, personal identification traits, serial numbers, permanent chip numbers, or any of a wide range of batch or individual identification. The key attributes of this type are specific, comparable, capturable, and controllable values.

Services and Limits

These data and logic establish areas of permissive Smart Card use and the limits or controls on those uses. These control functions come in several types. They may be narrow, singular, or specific. The controls may be multiple, conditional, and self-renewable. Or, they may be open-ended and based on external inputs or events. Service types, limits, geographic availability, usage locations, value limits, and total values available are key parameters. Other controls or limits may include account types, services, using devices, and the various financial transaction categories.

Amounts and limits may also be of several types, including total available, total used, or the new totals available on given dates. Also, totals in several categories include debits, credits, and cash. An added dimension may include system connection, protocols or routing permissible, network service options, interchange arrangements, and acceptable ranges of end-user initiated values.

Content

This category of Smart Card information covers the broadest, most flexible range. The content may be fixed format, variable format, or open-ended. The content categories include the following types:

Increments: Number, value, and periods of use.

Data: Type, format, amount, dates, referrals, error correction or detection codes, and small, self-contained data bases (e.g., hot card lists).

Authorizations: Location codes, control points, access codes, addresses, and routings.
Updating: Parts, dates, controls, inventories, and reorder points.
Logic: Application or jobs, security, procedural, and action options.
Human Factors: Messages, language designation, reconfiguration options or sequences, and preferred actions.
Currencies: Types, formats, exchange rates and limitations.
Instructions and programs: Using device programs, backup or alternative programs, initialization programs, closeout or summary programs, and other using unit programs (e.g., servicing instructions).
Others: Lists, capture areas, input/output controls, and product functional usage.

Security

Provisions for security information and logic varies widely. In some applications there may be no or minimal security in the form of checkable serial or identification numbers. At the other end of the spectrum are applications requiring a great deal of security. These applications may include the need for sufficient logic and data to permit stand-alone security control. The stand-alone techniques are based on user card content as augmented or compared with job card contents. The security content might include the following:

Access Codes: Values, usage periods, control points, control criteria, and acceptable personal or institutional access code values.
Authorization: Procedures, contacts, rules, sensitivity tests, and acceptable ranges.
Control values: Computation, message authentication check rules, formats, counts, and acceptable variables.
Encryption: Algorithms, keys, key exchanges, master keys, reformatting, key generation, and key management.

Audit and Journaling

This application function varies from no to full capture with historic event inquiry. The function may take several forms. One may be accessible only by the issuer. Another is accessible by the issuer and the Smart Card holder. In addition, service or security entries may be accessed with limited access control. Journaling entries may be accompanied by encrypted content controls. These are designed to prevent simulated or unauthorized duplicate entries. The journaling information may include the following:

Entries: Formats, dates, amounts, identifications, incomplete or incorrect entry tries, usage frequency, usage amounts in specified time periods, authorization codes, and captured values or status.

Journal mode: Number of entries, fixed or rewritable, and how the content may be used or accessed.

The journal entries have great potential value. They may be required to avoid a large number of Smart Card user inquiry requests. As users participate in a growing number of electronic transactions, they decrease their ability to keep adequate and complete records. Hence, an automatic inquiry facility can avoid a great deal of clerical effort to satisfy the card users. If a routine branch or home facility can be offered to let the Smart Card user do a self-service inquiry, then the issuer work load can be avoided. Also, pressure on the industry for more elaborate receipting of electronic transactions might be avoided.

CARD TYPES AND APPLICATION FUNCTIONS

In many industries there are three types of Smart Cards, the user, job, and device cards. Each type plays a different role, if it is required for an application. The user card concentrates on the variety of information relating to the user, to the user's capacity, transaction options, and record of events.

The job (or merchant's) card focuses on the application control and data capture. It may include application and security logic. It may be the delivery mechanism for transporting captured transaction data from the transaction site to the card issuer for record updating, billing, and payment reconciliation. The job card may also be used to deliver information to the application solution. This might include lists, loading an appropriate program logic, or even providing an accessible data base for application use. The technology implementation of the job card may also be quite different than the other cards. The user's card will usually be without its own power. The job card may have much more capacity and need its own power. The technology capabilities will be derived from the Smart Card designers' interpretation of the application functions to be provided by each card type.

The device card is used to control the information appliance being used as the Smart Card work station. The device Smart Card may have a broad range of capacities and a wide variety of technology options. In some cases it might not even be required. The device card brings the work station configuration, instructions, controls, and implementation specifics to the Smart Card application solution.

There are a variety and examples of application function contained in each Smart Card.

User: Identification, services and limits, appropriate content, security controls between the card user and the card issuer, and the journal.

Job: Identification, data capture, transaction control (e.g., a hot list), communications addresses or control points, and security controls between the card issuer and the job work station.

Device: Identification, configuration units and interconnection, communication addresses or access points, reconfiguration options, local diagnostics, and built-in recovery instructions.

The application card types are quite flexible within the given functional roles that each type plays. The application solution designer needs to make decisions based on the following criteria:

Which card types are required?

Which information will be included in each card type?

How is the application and security logic distributed between card types?

Which card carries how much data between the work station and the controlling system?

Which card carries how much control information and capabilities to the Smart Card transaction point?

What are the required card capacities and technologies?

What are the authorization levels, criteria and control points beyond the specified levels?

Which application modes are used? These include the following alternatives:

Stand-alone: This brings sufficient information and application facility to carry on a full transaction, within the established limits and controls, without further "remote" control.

Remote authorization/transfer: This requires that under limited or all-inclusive criteria, some or all of the transactions require control from a remote control point. The criteria may be amount-oriented, activity-oriented, or it may even be a random sampling.

Interactive on-line: This requires a dynamic movement of data and/or decision making between the transaction point and a remote host capability.

Understanding the answers to these questions and a selection of the application mode is important. The answers help to define the card type needed, their content and their role in the Smart Card transactions.

THE GENERIC SPECTRUM AND APPLICATION FUNCTION LEVELS

In the preceding chapter (see page 52) the concept of an opportunity matrix was introduced. The preliminary product function levels for the matrix were defined. The matrix role is to show the range of Smart Card opportunities. In Figure 5.1 the application functions are added to the hardware functions to complete the matrix boundaries.

Each intersection of the opportunity matrix represents a new generic Smart Card. (*Generic* means that each intersection is characteristic of a class of Smart Card application and configuration opportunities.) These functional combinations are each achieved by a Smart Card product function configuration and its preselected application functional capabilities. In the opportunity matrix the functions are shown cumulatively from left to right and from top to bottom. The generic opportunity matrix is best described as shown in Figure 5.2.

EXTENDING THE GENERIC EXAMPLES TO INDUSTRY OPPORTUNITIES

The opportunity matrix offers a view by industry areas of the generic user Smart Card applications. These are best portrayed on an industry-by-industry basis.

Financial Services Industry Opportunity Matrix

The financial services industry was used previously to identify and describe the financial transaction Smart Card. As the opportunity matrix in Figure 5.3 will demonstrate, there are many more opportunities for Smart Card usage in this industry.

What are the attributes of these Smart Card opportunities when compared to the labor and paper intensive conventional media solutions?

Increased capacity
Ease and response speed of the electrical connection

APPLICATION	PRODUCT FUNCTION		
	CARD		WORK STATIONS
FUNCTION	STORAGE	STORAGE + LOGIC	STORAGE + LOGIC + INPUT + OUTPUT
WRITE-ONCE DATE			
PLUS WRITE-ONCE JOURNAL			
PLUS READ/WRITE UPDATABLE			

Figure 5.1.

Control and security
Capture of the user-entered transaction data
Compact physical devices
Elimination of paper- and labor-intensive implementation
Support of new solutions based on new system attachments
Remove transactions from conventional branches

APPLICATION	PRODUCT FUNCTION		
	CARD		WORK STATIONS
FUNCTION	STORAGE	STORAGE + LOGIC	STORAGE + LOGIC + INPUT + OUTPUT
WRITE ONCE DATA	USAGE INCREMENT(S)	IDENTITY AND CONTROLLED ACCESS	A TRANSACTION WORK STATION FOR INQUIRY, DATA CAPTURE AND CONTROLLED ACCESS
PLUS WRITE ONCE JOURNAL	PLUS A USAGE RECORD	PLUS A USAGE & ACTION RECORD	PLUS A USAGE RECORD
PLUS READ/WRITE UPDATABLE	PLUS REUSABLE USAGE INCREMENTS	PLUS REUSABLE USAGE INCREMENTS	PLUS REUSABLE USAGE & INTERACTIVE TRANS

Figure 5.2.

The ability to use Smart Cards and interfacing devices in a remote or communications-based environment is particularly exciting. The money or cash type of Smart Card may be updatable through a remote telephone or television based work station. The updated Smart Card content could then be used at an electronic cash register as a cash

APPLICATION	FINANCE PRODUCT FUNCTION		
	CARD		WORK STATION
FUNCTION	STORAGE	STORAGE + LOGIC	STORAGE + LOGIC + INPUT + OUTPUT
WRITE-ONCE DATA	PAYMENT COUPONS	CHECK GUARANTEE CARD	INQUIRY STATUS WORK STATION
PLUS WRITE-ONCE JOURNAL	PLUS A PAYMENT JOURNAL	CREDIT OR DEBIT CARD	FINANCIAL TRANSACTOR WORK STATION
PLUS READ/WRITE UPDATABLE	PLUS AMOUNTS/ DATES UPDATABLE AND INTERACTIVE	MONEY OR CASH CARD	REMOTE OR STAND ALONE BANKING FACILITY WORK STATION

Figure 5.3.

substitute. This cash card mode would require an adequate security architecture for remote use in a stand-alone application. The "cash" card concept might significantly reduce the need for ATM cash dispensers. It would enhance remote or home banking by offering a new remote cash "issuing" facility.

Manufacturing Industry Opportunity Matrix

The manufacturing industries view the opportunity matrix differently, as shown in Figure 5.4.

In addition to the attributes identified for finance, the manufacturing opportunities extend into the following areas:

APPLICATION	MANUFACTURING PRODUCT FUNCTION		
	CARD		WORK STATION
FUNCTION	STORAGE	STORAGE + LOGIC	STORAGE + LOGIC + INPUT + OUTPUT
WRITE-ONCE DATA	INVENTORY WITHDRAWAL "COUPONS"	AREA ACCESS CONTROL	INQUIRY OR EXPEDITER WORK STATION
PLUS WRITE-ONCE JOURNAL	PLUS USAGE RECORDS	PLUS ACCESS, SERVICE, AND ATTENDANCE RECORDS	PAYMENT AND REQUISITION WORK STATION
PLUS READ/WRITE UPDATABLE	PLUS INVENTORY UPDATABLE AND INTERACTIVE	DYNAMIC ROUTING OR AREA ACCESS CONTROL	MANAGEMENT STAND-ALONE OR REMOTE WORK STATION

Figure 5.4.

Communications-based transactions, status, servicing, and control.
Mobile and flexible work stations with secure access and other Smart Card advantages.
Intelligent media usable at work stations, control points, and remote sites.

Public Sector Opportunity Matrix

The public sector or government usage area shows another interesting side of the Smart Card opportunity matrix, shown in Figure 5.5.

The public sector opportunity matrix adds a further set of application opportunities:

Public-carried, government-provided Smart Cards
Security and personal identification devices
Permanent movement records, such as passport use
Self-service-obtainable and -updatable licenses, documents, and permits.

Entertainment Industry Opportunity Matrix

The entertainment industry is in the process of being "married" to the communications industry. This industry, like the merchandising industry, is very responsive to hot sales items, if they know what is hot and can respond. Unlike the merchandising industry, the entertainment industry deals primary in information. Their current physical media (records, tapes, etc.) are produced and distributed by labor-intensive operations, making them vulnerable to serious delays in responding to fast market movements. This is resulting in great pressure to access and distribute the industry content by communications and remotely filled media, e.g. recording a cassette in several seconds at a store while the customer waits or pays for the transaction. (See Figure 5.6.)

Access rights, artists' rights, payment, security, distribution control, inventory control, and copy restrictions are essential elements of an entertainment industry solution. The Smart Card function has a natural role in this industry. Its information protocols with identification, security, prescribed services and limits, and usage control will facilitate the entertainment industry move to a communications-based environment. Smart Card devices may be used alone, in entertainment media (as an intelligent

APPLICATION	PUBLIC SECTOR PRODUCT FUNCTION		
	CARD		WORK STATION
FUNCTION	STORAGE	STORAGE + LOGIC	STORAGE + LOGIC + INPUT + OUTPUT
WRITE-ONCE DATA	SOCIAL PAYMENT "COUPONS"	IDENTIFICATION WITH PIN-BASED CONTROL	GOVERNMENT SERVICES INQUIRY STATION
PLUS WRITE-ONCE JOURNAL	LICENSE AND VIOLATIONS	PASSPORT WITH VISAS AND USAGE RECORDS	GOVERNMENT SERVICES TRANSACTOR
PLUS READ/WRITE UPDATABLE	UPDATABLE HEALTH/MEDICAL PROFILE	UPDATABLE PAY VOUCHER AND CONTROL	MANAGEMENT STAND-ALONE OR REMOTE WORK STATION

Figure 5.5.

video cassette) or within entertainment units (as a video player usage payment device).

AN APPLICATION EVOLUTION

In most industries the Smart Card will be introduced as a successor for an existing card or media device. For example, in financial services,

APPLICATION	ENTERTAINMENT PRODUCT FUNCTION		
	CARD		WORK STATION
FUNCTION	STORAGE	STORAGE + LOGIC	STORAGE + LOGIC + INPUT + OUTPUT
WRITE-ONCE DATA	USAGE ADMISSION	PERSONAL VALIDATED CONTROL	HOME DISTRIBUT'N & DESCRAMBLE CONTROL
PLUS WRITE-ONCE JOURNAL	PLUS "EVENT" RECORD	PLUS REMOTE RECORD	INTELLIGENT VIDEO & PORTABLE CASSETTE SERVICES ACCESS/RECORD
PLUS READ/WRITE UPDATABLE	PLUS RESERVATIONS & SEATS	PLUS UPDATABLE	REMOTE INTERACTIVE ACCESS

Figure 5.6.

it will follow the magnetic-stripe plastic card. What will be the application evolution? First, the Smart Card will be used as an extension of the active cards or media. It will be used for more storage capacity. Second, the Smart Card will introduce new functions. The PIN-based security feature will be used. Another new element will be the use of the ability to capture and journalize transactions within the smart card.

The third phase will be the introduction of new system attachment options such as the Store and Forward mode of work station operation. The fourth phase will be the addition of Smart Card function and features in the form of new work stations. These will be self-contained units with both work station and Smart Card functional capacity. The fifth stage will be the addition of Smart Card attributes to create work stations from existing units such as telephones and television sets. This will generate an even larger family of work station candidates. The sixth phase will be the introduction of Smart Card devices that did not exist previously. For example, an intelligent audio or video cassette for use in the entertainment industry. This unit could be used to dynamically receive new entertainment material via a communications based distribution system.

Application evolution will be available in incremental steps. These will enable earlier economic justification. The users will easily learn and accept change in manageable increments. Reasonable growth steps will encourage greater entrepreneural investment.

SUMMARY

These are a small selection of the multiple industry Smart Card opportunities. Further extension of the opportunity matrix to other industries will show a full range in each industry sector. This is an important message. The Smart Card is not an isolated event. Rather, it is a basic product with wide application. This characteristic is of significant value. It results in high use for each type of Smart Card. High volumes drive unit costs down. Lower costs provide more likelihood of economic justification of applications. (This will be discussed later in the *Economics* chapter, page 136.)

In summary, note that the combined logic and storage facilities of Smart Cards provide several levels of application function. Included are the write-once data storage, the write-once journal functions, and the write/read facility for updatable data. The application functions describe a broad set of opportunities when viewed with the range of functions. The Smart Card functions include storage, logic, and input/output facilities. Together these application and product function levels create the Smart Card opportunity matrix. This opportunity matrix portrays a full range of Smart Card media and work station product and application opportunities.

6 The Roles of Information Protocols

The Smart Cards will be carried by many users. They will be used in many devices. They will be accepted by many merchants, acquirers, and institutions. The Smart Cards will initiate many transactions, actions, processes, reports, interactive sessions, and will be subject to a wide variety of security system designs.

The issuers will have many expectations of the Smart Cards. They will desire a high degree of mobility, or the ability to use the Smart Card in an unlimited number of locations. They will expect to incur a minimum amount of system routings and communications "overhead" to use the Smart Card. Issuers will encourage decision making as close to the Smart Card usage point as possible with a reasonable level of confidence that the transaction is under control.

Smart Card usage will be analogous to the use of a coin telephone. The decision on coin validity is made in the telephone. The transaction identification, its destination, and the commitment to pay for the telephone call are all made through the phone dial entry. With a telephone credit card there is a reference to a central hot list and a checking algorithm on the card number entry. But no so with a coin telephone. You the user can go to any coin phone and carry on a telephone call with a minimum of bother and no reference to remote files or control points. Issuers want Smart Cards, and the work stations that accept them, to provide a similar set of immediate decision capabilities at the transaction location. This focuses on the issue of on-line control to a central data and activity information concentration point versus an off-line control concept using local logic and usage information carried within the Smart Card.

There will be other transaction needs to which the Smart Card will be expected to respond. As conventional communications line expenses explode, the issuers will seek alternative solutions to moving single messages on dedicated lines. The Smart Card that allows local decisions offers a way to eliminate communications messages for the major portion of transactions. The Smart Card will allow new system attachment options, including batch message transmission, which reduces the need for dedicated lines. However, the Smart Card will encourage message batches over dial-up facilities. This combination offers better economies of scale.

These diverse needs and expectations share one characteristic. They all require a prearranged set of information conventions. These are conventions that allow important operational questions to be answered at the right time, in the right place, with reasonable confidence in the results being correct. These conventions provide a basic set of information and decision making capabilities. They allow important actions and decisions to occur with assurance to the issuer and its remote control point and are called *information protocols*. Information protocols are the fundamental operational agreements to a common set of questions and responses. By providing these responses in the Smart Card, with its security capabilities, it offers a technique to achieve universal transaction capacity mobility, access, and usability.

WHAT ARE THE INFORMATION PROTOCOLS?

The Smart Card information protocols respond to the following questions:

Identification: Who are you? How do I talk to you? Who issued your card? To whom do I send the transaction details for billing and reconciliation?

Security: How do I know you are whom you say you are? How much has this card been used recently? To what value amount? Can I believe the value bearing elements of the transaction record?

Services and limits: What services do you want? What am I authorized to let you have? For how much? Under which conditions?

Audit and journal: How do we both capture the details for further processing and future reference? How do we protect the journal entries from tampering?

To these protocols must be added a variety of other Smart Card agreements and prior arrangements about the Smart Card itself. These agreements are generally referred to as *standards*. Standards describe the physical and electrical card and its basic identification. The protocols may also be standardized, but they refer to intra- or interindustry agreements. For example, the conventional financial transaction card has both national and international standards to describe the common information recorded on track two of the magnetic stripe. The initial standards area for Smart Cards covers the following information:

The physical card
The card contacts
The electronic signals
The interchange data

These standards are intended to provide a consistent basis of action. They are used in areas in which the Smart Card will be used by many people, in many applications, and among many acceptors.

What about application areas that are highly specialized or limited like access control facilities. These will need only to share a common manufacturer and users. The manufacturer will seek large volumes to obtain economy of scale. If achieved, the issuer will enjoy a better product price. However, the issuer will also insist on security safeguards or some set of unique traits. Hence, the specialized application areas will tend to be personalized to the particular product market.

Therefore, it is reasonable to conclude that the major demand for information protocols will coincide with the major cross-industry and cross-issuer environments.

INFORMATION PROTOCOL EXAMPLES

A better grasp of the significance of the information protocols is helped by considering several examples. The examples to be examined cover three areas. Consider the Smart Card opportunity matrix areas developed previously (see Figure 6.1).

Three typical Smart Card product opportunities are:

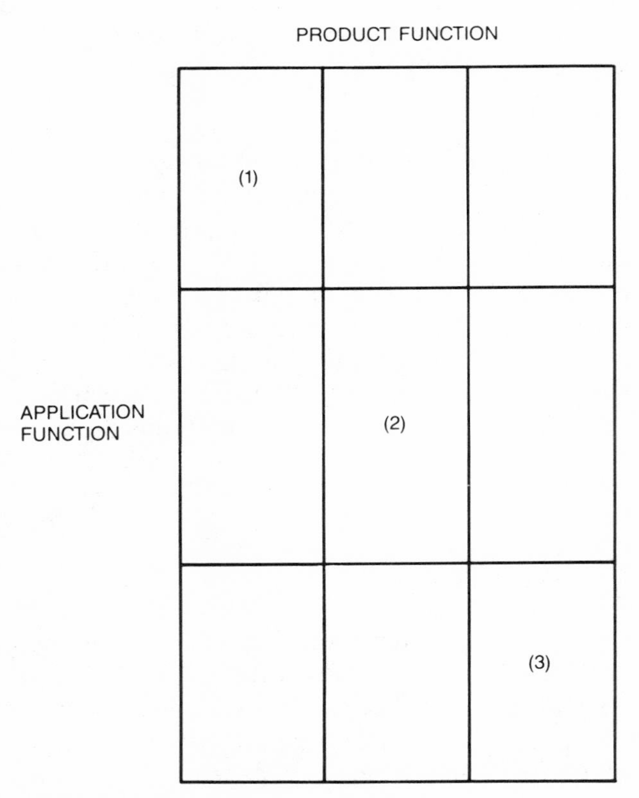

Figure 6.1.

1. An increment Smart Card: A government services payment card.
2. An interchange Smart Card: A videotex financial transaction card.
3. An interactive Smart Card work station: A remote management work station used for electronic purchases, vouchers receipt, and payments authorization and initiation.

AN INCREMENT SMART CARD

This type of Smart Card is issued in quantity, each containing a prespecified number of *increments* or units. Also, each card will generally have an issuer identification number, an issuer serial number, and possibly an issuer security code number. The increments may be consumed by the followed processes:

Inserting the Smart Card into an acceptor unit. This could be an attachment to an electronic cash register.
Validating or authorizing the Smart Card by reference to the job card and the Smart Card identification codes.
Consuming or using a number of increments according to a prescribed set of conditions. These could be "consumed" by destruction of stored increments.
Changing the card to reflect the reduction in available increments.

This Smart Card could also be a prepaid telephone or services payment card. It might offer increments of social payments such as food stamps or transportation payments.

What information protocols were used? Consider the following:

Identification: Is it our card? Is it an acceptable card?
Security: Is it our card code? Is the card number on the reported lost or stolen hot card list?
Services and limits: Are the required number of increments available?
Journal: None is provided, but the reduced number of available increments reflects card use.

These information protocols are dictated by the issuer. Normally, the issuer must make provision for the acceptor units to implement the protocols. If the acceptor units are unique, then they must be provided by the issuer. However, if the acceptors units are shared with other types of Smart Cards, then it may be possible to achieve better acceptor unit economics. For example, the acceptor could also be used for financial transaction Smart Cards.

THE INTERCHANGE SMART CARD

The interchange Smart Cards generally have a very broad issue volume and usage. In many cases it will be a worldwide mobility device. As such, the interchange Smart Cards will probably offer international standards requiring the following information protocols:

Identification: Issuer, user, interchange, country and industry of issue.
Security: The Smart Card will accept an externally provided personal identification number (a PIN). It will determine if the external PIN matches a PIN value placed in the card during its personalization process.

Services and limits: Available service types, current available funds, resettable available balances on prespecified dates (e.g., monthly), and countries whose cards may be used internationally, or areas of usage acceptance.

Journal: Usually provided with sufficient content to validate the merchant, date, and amount.

The interchange Smart Card may have a particularly challenging requirement. It may need to have sufficient information protocols defined for the card to be usable in an off-line or stand-alone environment. The interchange Smart Card would be usable only with access to the acceptor work station and its job Smart Card. That is, it is usable without access to the Smart Card issuer files at the time of the transaction. This card might be used for a financial transaction at a point-of-sale or through a telephone or videotex unit. This card will need to be usable at most transaction points under most circumstances. (The security design for this situation will be covered in a later chapter.)

AN INTERCHANGE SMART CARD EXAMPLE

A videotex unit is connected by a telephone line to a central videotex control facility. The central system distributes videotex frames on request. It also responds to user actions and requests. If the videotex user request relates to a merchant service, it is switched through the videotex control center to the third party. The third party is assumed to handle the request in its local facility (for example, the merchant's unit would handle the transaction). The videotex center might not have access to the issuer for authorization during transactions. Hence, the transaction process will rely on the merchant-based Smart Card facilities to maintain control, so a Smart Card interface unit would be added to the videotex. The videotex unit also provides a keyboard and display capability. The Smart Card functions will include the following:

Use and payment for the videotex services.

Selection and payment for products and services offered through the videotex service.

Selection and payment for banking and related services offered through the videotex service.

In each of the videotex-based functions the Smart Card sequence of action would start in a similar manner.

Insert card.
Select from the menu of actions offering the videotex services.
Enter the PIN value to "open" the card for a transaction. Security requires PIN acceptance by the Smart card.
Select the desired service or product from the menu.
The service and available funds limit must be accepted by the Smart Card and by the job card at the videotex center.
Transaction is journalized in both the user's Smart Card and the videotex central job card.

A similar sequence is followed in using the videotex system as an entry point to a third party service center. In both cases, following successful use of the information protocols, messages are prepared and transmitted for later posting to the appropriate records, bills, and balances. These messages may be batch transmitted from the receiving merchant's bank to the card issuer bank for billing. The messages convey transaction and payment specifics through the videotex facilities. Payments, orders, and transactions are carried to their conclusion through the videotex-connected communications facilities. In a large videotex center the job card may in fact be central logic processing and data capture on reels of magnetic tape that take transactions to their destination or to remote communication facilities, such as an automatic clearing house.

AGREEING ON THE INFORMATION PROTOCOLS

The primary challenge will be the effort needed to obtain the necessary multiple organizational agreements for creating the information protocols. This will be done first by one or two leading innovators in each industry. Then the evolved agreements will be taken to industry associations or national standards groups for consensus building and docu mentation.

The building process for creating the information protocols will evolve through the following steps:

Definitions
Agreements of early participants
Industry-level support
National standards development
Identifying and formalizing the national registration authorities

Encouraging participants to migrate to the industry or standard set of information protocols

Providing a process to accommodate further evolution and changes.

This process was followed in the development of the current financial transaction cards. There is currently in place an international registration authority for the identification numbers of the FTCs. The authority includes a standards-controlled registration and management process. If the existing process can be enlarged to accommodate the expanded information protocol opportunities of the interchange Smart Card, then the entire process can be continued and evolved. This would accelerate the growth and acceptance of the interchange Smart Cards.

Timing of the joint effort to formalize an information protocol is important. The effort must take place late enough to prevent premature freezing of these information conventions, however the results must be obtained early enough to create multiple user and vendor support for an extended set of issuers, preferably on a cross-industry basis.

AN INTERACTIVE WORK STATION EXAMPLE

This Smart Card is really a work station with Smart Card attributes. The Smart Card may be placed into a user card interface of the work station. The user card then carries on as if it were built in. A knee-top Japanese portable computer-based work station is currently demonstrating this Smart Card mode. There are a variety of operating modes for this type of work station.

Work station to work station: This is the electronic equivalent of a person-to-person transaction. The results are reported to the financial institution at the next occasion that the work station transmits a batch of events to the institution.

Stand-alone work station: This is an electronic equivalent to a session with your check book while paying bills. However, some of the bills may have been received previously by electronic mail. When a batch is completed it is transmitted to the issuer's financial institution during the next forwarding opportunity.

Work station to system: This is an electronic equivalent to mailing a batch of paid bills.

These added modes require an extension of the information protocols beyond those of the interchange Smart Card. For example, add the following modes:

Identification: Provide a permanent designation for the work station. This may coincide with the user's Smart Card identification.

Security: Add the ability to protect the work station storage and to protect the movement of data needed to support the work station modes described above.

Services and limits: These are enhanced to reflect the work station facilities. This information should be designed to speed communications to the work station such as use of the display (type and size) to select the desired services.

Journal: Enlarged to capture the external work station modes and connections.

WHY ARE THE INFORMATION PROTOCOLS IMPORTANT?

The prearranged information flow and responses of the information protocols achieves important results. They help to avoid some costly problems. These results start with recognition of the large number of participants involved in the Smart Card applications. Consider the participants with the financial transaction Smart Cards.

Card user/holder
Issuer
Supplier
Interface device or work station vendor
Acceptor or merchant
Acquirer or merchant's bank
Intermediaries such as a network provider
Communications facilities
Issue system
Billing system

Why are there so many participants? It is because the Smart Card opens a world of fully mobile financial transactions. Any time, any place, cross-industry, cross-country, and even cross-currencies. How does one achieve a uniformity of acceptance, ease of operation, and efficiency of performance across this wide variety of situations?

The answer is to prepare a set of agreements on the key questions and their responses. What are the key factors needed to proceed quickly to successfully conclude transactions with a minimum of external information needs? These agreements allow and require that all of the participants cooperate. These agreements allow the issuers to assemble systems from multiple vendors in geographically diverse locations.

As the number of Smart Cards grows and the usage locations increase, the combinations will be measured in the tens of millions. The use of information protocols will help to significantly reduce the on-line message traffic required to remotely authorize and confirm transactions. Imagine the need to validate your identification in the home office billing records before you are allowed to use a charge phone

	INFORMATION PROTOCOLS			
SMART CARD TYPE	IDENTIFICATION	SECURITY	SERVICES AND LIMITS	JOURNAL
INCREMENT	ISSUER	ID ONLY	FIXED LIMIT	NONE
INTERCHANGE	ISSUER AND USER	ID AND PIN	VARIABLE AND RESETTABLE	YES
INTERACTIVE	ISSUER, USER, WORK STATION	ID, PIN, AND VITAL INFO PROTECT	INTERACTIVE PLUS ON-LINE CONTROLS	YES AND WORK STATION MODES

Figure 6.2.

elsewhere in the world. The information protocols, when completed in the form described, avoid the need for messages seeking authorization for routine transactions: identification will be proven locally. Allowable services and usage limits are controllable at the transaction site through the Smart Cards, with a record of the event captured in both the user's and acceptor's Smart Cards.

The current French banking tests of financial transaction Smart Cards include information protocols of this type. Early experience in their use will provide guidance to the future development of these protocols. Thus far, however, there has not been a flow of information from these tests.

SUMMARY

By combining the previous descriptions, the range of information protocols by card type is arrived at. Figure 6.2 shows the variations in the information protocol parameters.

It would appear that information protocols for the entire range of Smart Cards are evolutionary. That is, as the Smart Card function increases, the information protocols expand and evolve similarly. If true, this is a most helpful characteristic. It means that the information protocol set can be built from a set of modules. The modules would add the necessary content and sophistication to match the functional evolution of the Smart Cards.

A key source of these information protocols will be the early Smart Card field tests, experiments, and applications. Hence, a better reporting of those results will facilitate recognition of the needed information protocol content and solutions.

7 The New System Attachment Modes

Smart Cards are intended to be used in work stations or have their function built into work stations. These product functions and the Smart Card facilities were described previously. But how is the work station attached to the remainder of the system? What is the "system"?

The dictionary defines "system" as an arrangement of parts into a unified whole. The work station is one of these parts, which when combined and connected provide a total solution. In fact, there are two systems. The first is the Smart Card system. These elements put the right card in the right user's hands with the right characteristics and personalization. The second is the Smart Card-based work station system. These elements deliver the right transaction in a secure manner to the responsible issuer to allow the proper posting and billing of the Smart Card-initiated transaction.

THE SMART CARD SYSTEM

The parts of the Smart Card system or the card life cycle cover all of the steps from the creation to the final retirement of the Smart Card:

Manufacturing
Issue

Personalization
Delivery
Initial use
Periodic content update
End of active usage
Retention for journal access
Destruction
Other elements
  Personalization changes
  Other content changes, e.g., services and limits
  Content change or updating, e.g., storage or logic

These parts are quite similar to those of a conventional financial transaction card. However, as a dynamic logic device, the Smart Card will be subject to a broader range of system interaction. For example, will the Smart Cards have their logic content corrected or upgraded without having it destroyed? How are the information protocols personalized, updated, or changed as required? How long is the Smart Card usable without being accessible from the issuing system?

Certainly some of the Smart Cards will be treated as the hand calculator of today. Once acquired, the user is stuck with the built-in logic features. The issued Smart Card will probably continue to perform without having access to later improvements or added capabilities. When the gap becomes sufficiently large, then it is appropriate to buy the improved follow-up device. The financial transaction Smart Card will have a planned life of only three years. Hence, there is an evolution that helps to respond to outdating.

Some Smart Cards will need to be supported with updating and changes during their entire intended life. It appears desirable to accommodate these changes in order to avoid an undue reissue expense over several years. Thus it is imperative that the Smart Card issuer have an explicit understanding of his Smart Card issue system and where it interacts with the Smart Card work station system. There will probably be several hundred Smart Cards issued for each work station. Therefore, careful thought needs to be given to making changes or updates through the Smart Card system or through the work station system. Another fact worth considering is that the work station system can probably be reached quicker with less effort through the job cards. These cards are being constantly recirculated. Hence, using a job-card-oriented change system may be more timely and economically sound.

THE SMART CARD–BASED WORK STATION SYSTEM

Work stations generally communicate with a system, which would include these elements:

Communications network
Communications front end
Application-based control processor
Issuer Smart-Card-related data base

There may be other system elements such as intermediate or interchange networks, regional processing centers, and distributed authorization centers with an authorizing data base.

The attachment of conventional financial transaction card work stations to a conventional system has several alternative system attachment modes. Whether they are on a dedicated line, a shared-line, or a dial-in facility, the most common attachment is on-line.

ON-LINE SYSTEM ATTACHMENT

On-line is a direct attachment from the work station to the application-based host processor and the issuer's data base. The transaction is then performed by interactive logic. There is application data flow between the two end points during the transaction. Performing an on-line transaction requires several key elements.

A communications line.
Appropriate logic in both the work station and the receiving end to control the line, move information, and control the process.
Application logic and data base interactions to perform the transaction during the attachment period.
Dedicated network equipment to respond to work stations, monitor performance, provide computer network management facilities, and to offer back-up capacity.
Host processor capacity to be used by work stations.

Thus, during the on-line interval, the work station has a prorated share of the work of the various system elements. This is not an inexpensive arrangement. However, with large transaction volumes and large work station populations, the shared expense per work station may be justifiable.

WHY THE ON-LINE SYSTEM?

Experience to date suggests that on-line systems attachment buys important results.

Direct control through central application logic and data base access.
Fast action with direct activity posting.
Updating records and billing without paper and labor expense or delays.
Reduced expense by replacing the physical movement of paper used in earlier approaches.
Immediate control of work stations and network status, availability, and operation.
Permits unattended customer self-service.
Offers immediate control of transactions based on reported lost and stolen cards and user account activity.

In addition to the advantages of supporting on-line work station operation, there are some disadvantages.

The high cost of communications.
Central or network outage, which prevents work station availability.
Network-based operations' vulnerability to security attacks, unless encryption or like-protection processes are introduced and used.
Poor system response time during peak activity periods.
Delays in on-line responses from interchange transactions requiring interconnection between systems.
High-volume transaction loads and large work station populations that require high-performance front ends with specialized processes and programs.
New magnetic-stripe financial transaction card frauds, which may negate on-line control. For example, low-cost cards could be used to counterfeit large volumes of legitimately issued cards in an application not requiring PINs. Even with an on-line system, this would result in at least three acceptable transactions per account number, per day, before the security sensitivity tests caused the transaction to be questioned.

In the United States, on-line has been the desired mode, particularly for high-volume point-of-sale locations. Intermediate volume locations have used dial-up facilities. Low-volume locations have used a manual hot list look-up and phone authorizations for high-value transactions.

THE NEW SYSTEM ATTACHMENT ALTERNATIVES: REMOTE AND STORE AND FORWARD

Several new important options are opened by the advent of the Smart Card with internal logic and security architecture. These options are encouraging the examination of new system attachment modes, particularly off-line modes and distributed processing. This examination process may be accelerated by a number of important marketplace events:

1. Communication expenses are increasing rapidly. This is particularly so with local dedicated lines within a 300-mile radius, in most parts of the United States. Many of the target locations for work stations have a very low geographic density and thus do not have access to new communications technology alternatives. Nor can their traffic be concentrated to reduce line expense.
2. Conventional plastic card frauds are increasing at a very high rate. The planned responses are, at best, stopgap. It appears logical to expect that other frauds will start to accelerate for magnetic-stripe cards. Hence, on-line solutions may not be adequate for mass card counterfeiting, communication-line tapping, or network attacks such as "spoofing." (Spoofing involves inserting an illegal device into a network to give work stations an apparently legitimate authorizing response.)
3. Expenses are also increasing because of the interchange arrangements, interstate banking, and like changes, which are increasing the geographic distances and switching complexities involved in conventional card authorization.
4. The rapid growth of integrated circuit chip capacity suggests that the expense of local card authorization and application control is decreasing. Hence, a local capability with adequate security application control may become economically competitive with on-line solutions.
5. Telephone-based equipment and services are being rapidly modernized and expanded. There will be new phone devices accepting cards. There will be new work stations introduced, for example, videotex, which will require card-based control and application services. There will be new network services, like digital data packet switching, which will make transaction routing through telephone-based equipment very attractive economically when there is adequate usage to justify installation.

These changes demand an examination of new system attachment alternatives and operating modes. There are two attachment modes in particular that look attractive with Smart Card usage and work stations, *remote* and *store and forward*.

THE REMOTE SMART CARD WORK STATION

The remote system attachment mode uses telephone or cable communications facilities, terminals and work stations with Smart Card interface capabilities. These alternatives will include a variety of units.

Conventional public telephone with a Smart Card interface.
Videotex unit with Smart Card interface and keyboard
Electronic directory unit with Smart Card interface, display, and keyboard.
Display-type public telephone with a Smart Card interface.

A Smart Card transaction performed through a conventional telephone might take several phases. Consider using a financial transaction Smart Card to purchase an article from a merchant with a remote system attached to a conventional telephone:

Place the Smart Card into its receptacle.
Dial the merchant.
Enter your PIN through the telephone touch tone pad to enable the Smart Card to pay for the telephone call. The Smart Card could also indicate the user's language preference.
Reach the merchant, discuss the transaction, and agree to the charges for a placed order.
The merchant enters the transaction amount into his work station and its amount is displayed to the Smart Card user for confirmation.
The customer reenters his PIN number to open the Smart Card to accept the merchant's transaction.
A unit at the merchant's telephone contains his job card. The merchant's unit authorizes the transaction based on Smart Card-provided data and the job card controls. The merchant's unit captures the transaction data for later posting and billing through the issuer's institution. Thus the Smart Card is on-line to the merchant.
The customer removes his Smart Card and a receipt is provided with the merchandise delivery. The customer may access later his Smart Card journal entries for inquiry response.
At the end of the day or period, the merchant takes or transmits

his job card content to his financial institution. They capture the transactions, forward the data to the Smart Card issuing institutions for posting and billing, the job card is cleared, and its hot list is updated.

The advent of remote financial system attachment significantly enlarges the number of work stations and locations through which a financial transaction Smart Card may be used.

THE ADVANTAGES OF REMOTE SYSTEM ATTACHMENT

The remote environment is the public switched network and its shared economics. It is a broadly available facility. The financial transaction Smart Card users will be quite comfortable with the use of remote facilities, especially within a few years as a result of the emerging role of conventional magnetic-stripe cards as telephone credit cards. The remote system attachment mode provides the following advantages:

Shared work station and network economics
Broad transaction mobility
Elimination of local paper- and labor-intensive transactions
Movement of transactions away from conventional data system facilities
Increased marketing options available to merchants
On-line control for telephone-based facilities (e.g., videotex services)

The critical issue in undertaking remote system attachment is the ability to maintain control and security. Can the same or better control be exercised with the Smart Card? In the financial institution Smart Card two elements of control are added: introduction of PIN-based access control to the card content and its use, and access to the in-card amount and usage controls.

However, two elements are changed significantly. First, the centralized controls for non-telephone company cards against wholesale card fraud by examining consolidated issuer-based activity records is dropped. Second, the local hot list in the accepting system will not be as complete or as current as an issuer-consolidated record of activity for that account.

These issues will be discussed more in the chapter on Security. However, the overwhelming advantage of handling financial transaction Smart Cards transactions by remote system attachment suggests that an adequate security solution will be set. In fact, security is really a compromise between risk and reward. Hence, several compromises are available that seem to provide an adequate answer. The choice will be made by the key players—the issuers.

THE STORE AND FORWARD SMART CARD WORK STATION

The store and forward systems attachment mode is partially on-line. This mode handles financial transaction Smart Card transactions in a work station that is not routinely attached to the central controlling system or through a communications facility. Rather it relies on the user's card content, the merchant's job card content, and the work station logic. Exception transactions, those over a set limit, result in a dial-connected, on-line authorization transaction. Completed transactions are accumulated during work station usage. At the end of the day or period, a batch of stored and completed transactions are forwarded to the merchant's or acquirer financial institution. Hence, the description as a store and forward systems attachment mode.

The routine transactions are treated in the store and forward mode. These are generally the largest percentage of the transaction volumes. The difficult, the exception, and the rejected transaction attempts are all treated as on-line activity. As a minimum, this system attachment mode will materially assist in migrating the work station away from requiring a dedicated communications line.

The store and forward mode therefore offers these advantages:

Utilizes dial-up facilities when an exception basis is detected.

Avoids the need for dedicated lines and shared dedicated network facilities.

Reduces system capacity needed to handle on-line transaction processing.

Replaces paper- and labor-intensive mode of operation for off-line imprinting units.

The Smart Card user and merchant does not need to know, and in fact may not know, if the dial-up, on-line mode is in use. This is a psychological assist in resisting bad card passing.

The Smart Card retains a permanent journal of transactions. This allows the user to handle his own inquiry requests for Smart Card activity questions.

The Smart Card logic allows control of transactions based on their internally provided dollar value, usage frequency, or total value.

A printed hot list is no longer needed and is replaced by the in-card balances and controls and the merchant's cassette containing the hot list.

The clerk's time and time required for manual check writing is significantly reduced.

In a period of rapidly increasing communications expenses, the store and forward system attachment mode appears to be desirable. It should reduce several expenses. However, these savings may be offset by the added cost of Smart Card work station enhancements. An important economic fact to note is the ratio of Smart Cards to work stations. In general, there are many more cards than work stations, possibly several hundred to one. Hence, a dollar of increased cost per user card might be more expensive than a few hundred dollars charge for improved performance per work station and its job Smart Card.

THE NEW SYSTEM ATTACHMENT MODE APPLICATIONS

The Smart Card with its self-contained information protocols fits well into the remote and store and forward modes. These are realized on new access facilities:

Videotex
Portable work stations
Electronic directories, catalogs, want ads, yellow pages, and teleshopping
Medical diagnostic units and displays
Equipment servicing
Remote processing unit access

Each of these units will benefit from Smart Card capabilities and the new system attachment modes. The information protocols are the key ingredient in these emerging scenarios. The identification, security, and services controls set the stage for these transactions.

How are these transactions implemented? Consider the following sequence of events.

Establish a working relation between the user's Smart Card, the work station job Smart Card, and the work station. The work station may also have its own device card.

- Insert the cards.
- Enter the transaction amount.
- Confirm the user's acceptance of the transaction and its cost by his PIN entry.

Enable the transaction to occur between the work station elements.

- Apply the application logic.
- Capture the appropriate data.

Perform the required authorization criteria.
Prepare the journal entries.
Prepare the job card entries.

Report the store and forward results to the central data gathering point.

Dial the access point.
Transmit the batch of captured transactions.
Receive an updated job card content.

These processes enlarge the system attachment options available. To the existing on-line mode has been added store and forward where batch efficiencies are available and where the bulk of routine transactions can be handled. This is a valuable system expansion opportunity. It enables locations that do not justify full on-line communications to be added to the electronic system. However, data capture will be delayed to the batch forwarding time.

VISA Super Smart Card (Figure 7.1)

For off-line environments an added facility is needed to use the Smart Card content without a system attachment. One approach is to build an off-line authorizer into the Smart Card. Consider the Super Smart Card, a card concept aimed at 1988 test cards and a 1990 business use for off-line usage. The Super Smart Card is four cards in one. First, an embossed card, metallic, with raised embossing. Second, a magnetic "striped" card WITHOUT a stripe! It has a stripe simulator that emits the equivalent magnetic fields. To use this feature in a slot-type card reader will require marking a card placement area on the slot, and putting the card in place so that the magnetic-stripe simulator output reaches the stripe reader head. If MCAS (track 1) is also in use, then a two-part simulator (tracks 1 and 2, high density and different position from the card edge), will be required. The simulated stripe will probably NOT be usable in any card reader that moves the card past the magnetic read head—as in an ATM card reader.

The third card is a Smart Card. The contacts location is below the stripe. The signals will be consistent with the appropriate financial transaction Smart Card standards. The fourth card is a new off-line authorizer. It consists of a crystal display (16 characters), a 20-key keyboard including a 12-key, touch-tone equivalent numeric pad and eight functional keys, 64K bits of storage, and a built-in battery power supply.

The Authorizer is "opened" with the PIN entered by the customer. The merchant, at an off-line location, enters the transaction amount

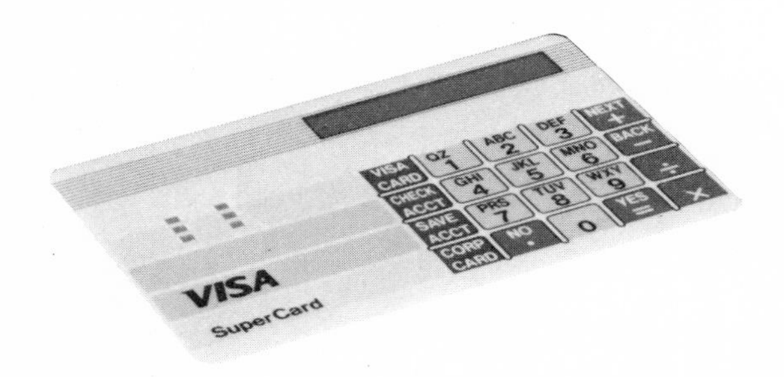

Figure 7.1. The VISA Super Smart Card *(Courtesy VISA International, San Mateo, California)*

through the Authorizer keyboard. If the transaction is within the available balance, the Authorizer display shows "approved" and an authorizing number that is copied manually on the imprinted forms set by the merchant. If the transaction value is not within the available limits of the card, the displayed message tells the merchant to call the authorization center. Each month, on the monthly billing statement, a 6-digit number is included. When the 6-digit number is entered into the Authorizer keyboard by the card holder, the available credit line or account balance is updated for Authorizer transactions during the following period.

SUMMARY

The Smart Card allows transactions to occur at work stations and to be communicated through new system attachment alternatives. These alternatives are *remote* system attachment through the public switched network and *store and forward* system attachment for batch transmission after a number of transactions are completed.

The Smart Card work station system consists of those elements needed to communicate with and support the operation of the Smart Card-based work station in one of the new attachment modes. The work station must establish a clear relationship between the user's card, the issuer's control (perhaps in a job card), and work station control (perhaps in a device card). Once this is done, a transaction can be processed and security established and checked. Later, the results need to be documented, recorded, and communicated.

The Smart Card life cycle system consists of the steps necessary to create, deliver, use, and retire Smart Cards. This system needs to be able to accommodate change, evolution, and improvements even though a substantial number of cards are issued and in use. The Smart Card design and those of other system components will never be frozen. The card life cycle system must be designed and managed with this fact in mind.

8 Field Tests

Field tests in the early life of new products serve many purposes. They may range from seeking early experiences and user acceptance confirmation to pure public relations. The public relations may be aimed at users, vendors, industry associations, and standards groups. In this chapter we will focus on the nature and intent of the current field tests.

Financial transaction Smart Card field tests currently appear to be focused on encouraging market usage and validating point-of-sale work station use. Other important issues such as card survivability, security architecture validation, product requirements confirmation, economic justification, and system attachment feasibility appear to have been laboratory tested during the development phase, but must also pass the test of everyday use.

EARLY FINANCIAL TRANSACTION CARD TESTS

The first large-scale public tests of FTCs with magnetic stripes occurred in early 1970. The test followed a large amount of laboratory testing. It involved issuing 15,000 magnetic-stripe airline self-ticketing at Chicago's O'Hare airport. The test objectives included card survivability, market acceptance and card usage, human factors confirmation, and self-service design confirmation.

There were several unexpected results, like stripes not adhering under certain circumstances and several unsuccessful attempts to penetrate the security design. However, the overwhelming acceptance and use of the cards and the repeated use of the self-service facilities were the important results. In fact, some passengers walked halfway across O'Hare from airline counters with long ticket queues, bought a ticket by self-service, and returned in less time than waiting in line at the counter. This type of acceptance and user activity confirmed the market interest in self-service. Ironically, the banking industry then proceeded to move to self-service implementation, rather than the airlines. Also, the laboratory testing of the magnetic-stripe card survivability was confirmed by test results.

An important public relations result was the positive influence of the test results on the creation of an American National Standard for the magnetic stripe financial transaction card. Vendors, suppliers, industry groups, and related participants could see the new cards at work. The demonstrated users' acceptance was most helpful. As a result, the American National and International Standards evolution was encouraged, and achieved in the subsequent few years.

THE SMART CARD FIELD TESTS

There are three types of tests currently underway. They cover: (1) banking and financial services; (2) telephone and dial-network delivered transactions; and (3) miscellaneous military, medical benefits, and access control tests. The largest tests are underway in France.

The French Test Environment

The French government appears to have embarked on a course of action to:

Achieve world leadership in electronic payment systems and products.
Demonstrate this leadership through a series of extensive tests in France, which will lead to international market acceptance and use.
Create an exportable set of related technology products.

Smart Cards are one of the products being tested. The Smart Card tests in France have been directed to two major market areas. One is banking use at point-of-sale with a store and forward system attachment mode for work stations. The other is telephone-based Smart Card tests using the remote system attachment mode for public-switched network devices. In both market areas there appear to be important application and economic benefits based on the current French solutions and costs.

EARLY FRENCH SMART CARD TESTING

There has been a good deal of Smart Card physical technical testing information developed by the French National Communications Laboratory. Their reports have been given to the International Standards Organization working group on Integrated Circuit (Smart) Cards (ISO 97/17/4). They are available through the United States participants (ANSC X3B10.1). There is, however, one note of caution. These technical reports appear to have removed any proprietary information relating to any of the French vendor cards. Also, the test cards are different designs from each of three vendors. These differ from the consensus design that is being drafted into the international draft standards. There is no Smart Card chip design, application description, information, or Smart Card content layout in these reports.

These laboratory test results portray an extensive ability of the French Smart Cards to withstand a broad array of physical tests, environments, and contamination. But there are some missing elements. For example, the tests did not measure or attempt to discover the heat dissipation from imbedded chips or the impact of chip dissipation heat on the magnetic stripe or the physical card. This has been handled in the proposed physical card standard by specifying a maximum heat dissipation level. Also, the proposed standard location for chip contacts and, hence, probably the chip itself has been moved away from the stripe location as initially tested in France.

Overall, there is still a lack of significant field test reports. Nothing has been reported on card survivability, security architecture success, human factors experience, work station functional performance, transaction volumes, or economic justification findings. The vendors involved probably consider all of these issues to be proprietary. In addition, any test of this type will produce some unexpected results—a portion of which are probably negative. Results that don't fit the scenario are sometimes not too broadly reported. They are too valuable!

The French PTT Communications Smart Card Tests

A population of 2500 interactive Teletel videotex units were installed by the national phone system (the PTT) in 1982-83. Their use is free. About 300 of these work stations have a Smart Card interface. Each of these 300 units is tied to one of 250 postal (PTT) financial services accounts or to one of 50 regional bank accounts. The test Smart Cards were reported to cost $40 each (but that may have been at the old exchange rate of 4 French francs per dollar; at the current exchange rates the card would cost $16 each).

There are three Smart Card usage modes used in the tests.

1. Payment for telephone calls: The Smart Card is used as a telephone credit card. It may be used in a new public phone or videotex unit or in an electronic directory unit. The latter are keyboard-display units intended to replace printed directories for white and yellow page services. These are called Minitel units.
2. Payment for videotex access fees: The Smart Card is used as a telephone credit card but the charges are based on the requested videotex services.
3. Use as a banking card: The Smart Card is used to make payments through merchants or access a banking account and services. The Smart Card has a journal capacity of up to 80 transactions. This limits the maximum usage of the card. The Smart Card also provides for an automatic available balance refreshing each month to its maximum payment capacity.

These applications are essentially the same as the remote system attachment mode applications as described previously. The application flow starts with Smart Card insertion into the telephone, into the electronic directory, or into the videotex unit. The available banking functions include the following:

Balance inquiry
Recent transaction details
Orders for statements and checks
Account-to-account transfers in the same bank
Telepayment by an "electronic checkbook" using the Smart Card as debit or direct charge card

In dealing with a remote merchant, a payment message is sent to the merchant after customer acceptance of the amount. This is initiated with a PIN entry from the customer. The merchant forwards a batch of captured transactions to his bank, and from there the transactions are sent through a clearing house to the Smart Card issuer's institution. The latter method of operation from the merchant to his bank is similar to the store and forward system attachment mode. Few results have been disclosed from any of the French communications services or banking tests.

For payments of phone calls, the French PTT is introducing a new public phone that has a built-in Smart Card interface. These telephones will accept three types of Smart Card.

A prepaid telephone payment card: This card has a preset number of telephone call payment units. The units are consumed at a variable rate based on the telephone toll charges.

A telephone charge card: This Smart Card acts as a credit card. Telephone charges are billed through the telephone system billing process.

A bank Smart Card: This Smart Card acts as a banking card. Charges will be paid in a manner pre-established by the bank. The card may be a credit, a debit, or a cash payment card. The telephone charges will appear on the banking statement to reflect Smart Card activity.

Early reports based on 1981 experiences with the Smart Card used before that used with French public telephones described the justification for these Smart Card applications. In 1981 it was reported that 15 percent of phone call tariffs were lost because of forms of telephone charging frauds. This is consistent with evidence that coin thefts from telephones and coin collection expense are very large all over the world. The Italian telephone system, which has also described a new Smart Card designed as a prepaid telephone payment card, has reported that two thirds of coin telephone unit costs relate to the coin mechanism, coin protection, and antitheft provisions. Hence, a Smart Card approach is quite attractive. It offers a non-coin-based payment system, eliminates coin collection, and provides a more secure direct payment and direct billing system.

The Smart Card-based telephone does not have a coin mechanism, but it does have four "function" buttons to the right of the key pad. There is a display and a Smart Card receptacle. These features allow for remote financial transactions. The unit is reported as accepting three types of Smart Cards: prepaid telephone calls, telephone credit cards, and financial transaction cards.

The remaining balance of the prepaid card is probably displayed when initially inserted. Future usage of card-based phones such as this unit are a function of the number of units to be installed and the using population of cards available in the same geographic area.

Another French telephone test is the application of Smart Cards with some of the new electronic directory or Minitel terminals. These keyboard-equipped display terminals are intended to replace printed telephone directories. One of the intended applications is the presentation of catalogs or related types of "yellow pages." Hence, this work station provides a form of teleshopping with Smart Card access.

The Future of the Communications Smart Card

Ten thousand Smart Card pay telephones were installed in France in 1984 with 400,000 Smart Cards issued for telephone use. It has been reported in the press that the French PTT has ordered and has issued a total of 1.4 million financial transaction Smart Cards in 1984. These cards are for all of the new PTT services plus post office-operated banking and savings activities. A volume of 10 million has been discussed for 1986 issue. At the same time, a commitment to more telephone-based work stations of all types has been announced. Specifically, a large-scale installation of public phones with a Smart Card interface is committed and started in 1985 with 30,000 units, along with a large-scale electronic directory installation program (Minitels) with volumes in the millions. Expansion of the videotex installations has also been announced. There were about 100,000 units installed by mid-1983; 600,000 units are reported under production, with an added 500,000 units ordered early in 1984. Minitels are being installed at the rate of 500 per day. In September 1983 there were 35,000 installed units in homes plus 12,000 units in a variety of businesses as Professional Teletel Services. The 1983 total was over 100,000 units. A population of over 3 million units was achieved in early 1986.

Despite the absence of results reporting, the PTTs of other countries are about to seek bids for Smart Card field tests. The German PTT is reported as seeking an early 1985 test.

THE FRENCH BANKING INDUSTRY SMART CARD TESTS

These tests have been focusing on a point-of-sale application. Sixty percent of the checks written in France are written at a point-of-sale. Also, check clearing (returning the check to the issuer's bank) is reported to cost four times the cost in the United States. Banks claim that it is not possible to pass these costs to the check users. In addition, in France, the costs of dedicated communication lines are about three to five times the expense of those in the United States (in 1983). Put all these factors together, and the concept of Smart Card use at a point-of-sale in a store and forward mode becomes very attractive and, possibly, very economically justified (Figure 8.1). Consider the following justification factors:

Check and check-clearing expenses are eliminated.

The store and forward system attachment mode avoids the need for dedicated lines.

The built-in Smart Card payment and authorization process decreases the checkout-stand transaction time.

The manual authorization process and support, and printed hot lists, are eliminated.

Three tests have been installed. Each is in a different city of France (Blois, Caen, and Lyon), with equipment from a different vendor (Cii Honeywell Bull, Philips of Holland, and Flonic Sclumberger), and a different Smart Card design. Each test has about 200 point-of-sale work stations installed in merchant locations. The entire test was planned to last from mid-1982 to the end of 1984. The Smart card population objectives were 25,000 in Blois, 30,000 in Caen, and 50,000 cards in Lyon for the three tests.

Although a total card population of about 120,000 cards were manufactured by mid-1982, only 60,000 cards were in customers' hands by the third quarter of 1983. In addition, it was reported that these cards were experiencing a low-transaction volume. The stated reason for the shortfall in customer acceptance has been alternately described as due to poor marketing and to a "float" difference from credit cards with a 25-day float. The Smart Cards in the test were used as electronic checks and therefore result in transaction posting within three days. Most bank customers react as you and I would, to seek the better deal in the conventional credit card float and don't really care that a Smart Card test is under way. Hence, the low Smart Card acceptance and usage.

As previously stated, there has been no significant disclosure of any results from the banking industry Smart Card tests. There have been rumors of some technical problems. Generally, the vendors have insisted that the tests have gone according to plan, at least technically.

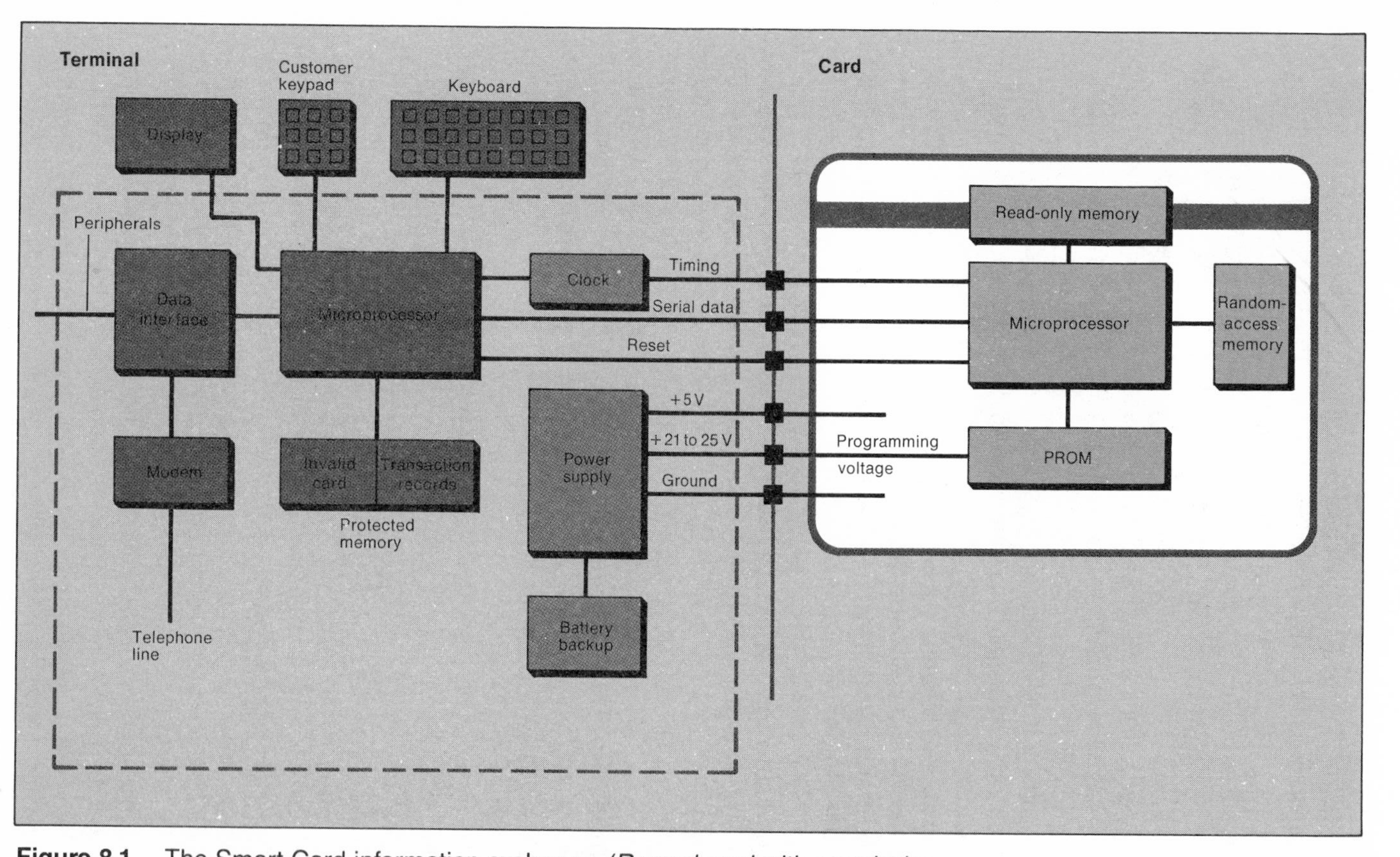

Figure 8.1. The Smart Card information exchange. (*Reproduced with permission from Steven B. Weinstein: "Smart Credit Cards: The Answer to Cashless Shopping,"* IEEE Spectrum, *February, 1984. Copyright © 1984, IEEE.*)

FUTURE PERSPECTIVES FOR THE FRENCH BANKING SMART CARD

In late 1983 an important test followup decision was announced. The French affiliate of VISA, Carte Bleu, indicated that up to 12,000 Smart Cards with credit card capability were issued in one of the three test cities in the first quarter of 1984. Because it is a credit card with delayed payment, this card issue would probably overcome the float resistance of customers in accepting and using the previous check replacement test cards.

The initial French banking tests are operated in an application mode called an "electronic checkbook." The terminals are using a store and forward mode of system attachment. However, two additional application modes have been described for future use and testing. One is that of a credit card. The recent Carte Bleu test announcement would appear to start the credit card application. The other mode is called an "electronic wallet." This treats the Smart Card as a cash-only Smart Card version. This mode requires segregating cash in an account for limited access, only through use of the Smart Card. This is an important difference. It provides a means of servicing customers who may not want or do not qualify for either credit or debit access. Also, with the added 100 percent authorization and control of the Smart Card, the issuing bank is assured that the use of this card can not exceed the available funds. If this proves implementable, it opens an exciting application area with a genuine electronic cash medium.

With the three application modes, credit-debit-cash, the entire spectrum of banking applications will be explored with the Smart Card as the test vehicle. These modes will be discussed further in the chapter on economics (see page 136).

OTHER SMART CARD TESTS

A variety of other Smart Card tests are in process in several parts of the world.

1. United States military data tags: These are *lower-left opportunity matrix* (see page 64) Smart Cards with storage and in the shape of a soldier's dog tag. The application is battle field logistics control and reporting of personnel and their skills.
2. United States military dependents medical services authorization and payment: These are financial transaction or *center oppor-*

tunity matrix Smart Cards. They are used to identify recipients and to pay for their services.

3. Italian telephone payment card: These storage-only, *upper-left opportunity matrix* Smart Cards that are especially designed to be tested as a multiple-increment telephone payment card. The card is intended to be purchased at a local point and in advance of placing the calls. The increments are then consumed as phone calls are made. The Smart Card receiving phone has no coin mechanism. The phone does display the residual value of the card. The phone also provides a brief interval to remove a consumed card and to replace it with a value card during a phone call. The same facility has been described for the French PTT public telephones.
4. Japanese program load card: This appears to be a *left-center opportunity matrix* Smart Card. It is used to load a program of up to 3000 characters into a knee-top personal portable computer and work station. These cards have been demonstrated with 0.9 and 1.5 times the thickness of the standard financial transaction card. The demonstrator stated that the thicker card exhibited more reliability.
5. Japanese portable data base Smart Card: This appears to be a self-powered *lower-left opportunity matrix* Smart Card. This card was shown in the same demonstration as the preceding Smart Card. These cards are about 6 times the thickness of the financial transaction card. The Smart Card contains 13,000 characters of data base. It is used in the knee-top computer mentioned above. This appears to be a job card. The card contains battery-driven, high-density, read and rewritable chip storage.
6. French medical applications: Three tests are planned. In Blois, 8000 cards will be issued for medical information and vaccination data. Interface devices will be available to physicians and at schools. Two added pilots will distribute cards for chronically ill patients and to outpatients with need for frequent medical care.
7. Germany contactless card: A Smart Card without contacts is being demonstrated. It uses a very high frequency antenna in the Smart card to receive power and to handle data flow. The International Standards group on Integrated Circuit Cards has stated an interest in extending the standards modules to include a contactless version when a written standards proposal is received.
8. United States key-type cards: These units are commercially available. They provide both storage and logic and are used for

a variety of access control, servicing, and industrial applications (Figures 8.2 and 8.3).

There are a large number of Smart Card configuration alternatives. The Smart Card opportunity matrix portrays the range of options and configurations possible. Other Smart Cards are under development and this includes a *lower-right opportunity matrix* complete personal computer in a card size physical package, which should be introduced and demonstrated soon. There is also some investment being made in conventional magnetic-stripe financial transaction cards and other technological alternatives. These will be discussed in the chapters on alternative technology (page 145) and magnetic stripe evolution (page 165).

An important opportunity for future field tests is to complete the definition and content of the opportunity matrix. Early testing continues to focus on the financial transaction Smart Card for obvious market size reasons. Clearly this is an important market opportunity. However, inspection of the opportunity matrix indicates only limited testing of the Smart Card packaged in its own work station and with its own input/output. Also, there is a need to test the Smart Cards in an interactive mode or *lower-third opportunity matrix*. These configurations will evolve from future application definitions and field tests.

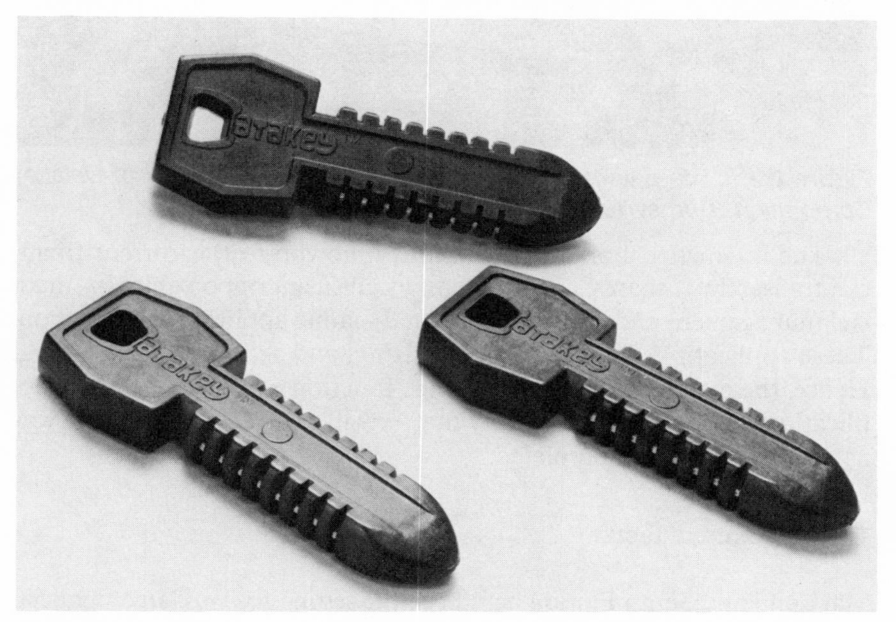

Figure 8.2. The datakey Smart Card. (*Courtesy of Datakey Corporation, Burnsville, MN.*)

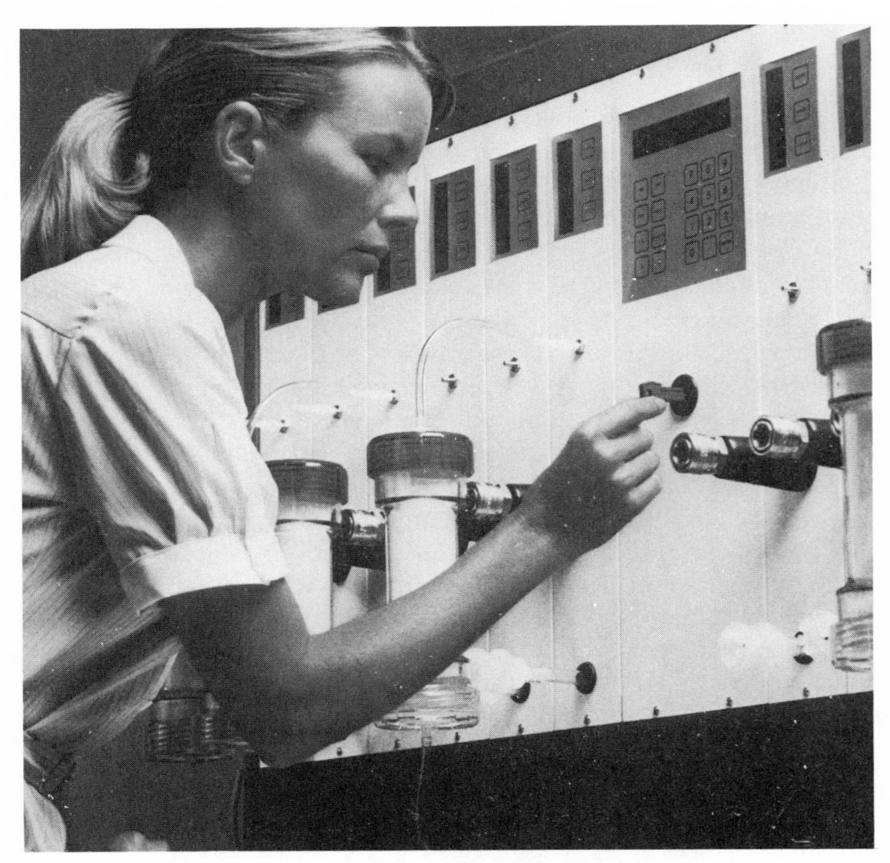

Figure 8.3. The datakey in a medical application. (*Courtesy of Datakey Corporation, Burnsville, MN.*)

The interactive storage is a natural followup to the current financial transaction Smart Card. It opens a significant opportunity in financial management and control based on dynamic application interaction. These applications require receiving, using, and acting on current data. Hence, the dynamic read/write storage function complements these application needs. The opportunity matrix will continue to point the way as current tests are completed.

The Mastercard Tests

Maryland and South Florida have been the setting for two large financial transaction Smart Card tests. Forty thousand cards plus 200 POS, split between the locations, comprised the bank test started in 1985. Ten banks participated, and by 1986 normal card usage was achieved. The plan is to extend the test another year and, then, to replace the test gear by pilot production equipment manufactured to Mastercard specifications.

The Japanese Tests

Every major bank, retailer, equipment vendor, card supplier, and government agency involved in transactions is involved in one of 40 tests by late 1986, and these are expected to continue for at least another year. By then a new national set of Smart Card standards will have been formulated. They will be used for the test follow-on equipment procurement. There were at least 25 vendors of Smart Card application products demonstrating almost 100 applications at the recent ICC International exhibition in Tokyo. That should stimulate added testing.

U.S. Department of Agriculture Tests

A 2,700 farmer, 60-farm commodity buying point test was completed in late 1986. It will expand to a full 100,000 card and 1,000 reader system to supervise the purchase of allocated quotas for the peanut crop. The Smart Card solution offers decentralized control authorization with significant paper processing reduction while creating a nationwide online batch capture and collection capability. The Smart Card solution combines marketing point control and flexibility with paper work simplification.

American Telephone and Telegraph Company

Testing a contactless card in display telephones in several cities. Also, this card is being actively marketed as a Financial Transaction Card and as a military data tag for secure access and personnel card. This card uses capacitive signal coupling and inductive power.

SUMMARY

Tests are becoming "nonstop" entries into production equipment-based systems. The nature of the tests are shifting from equipment and survivability validation to new business rules and economic model validation. These latter results promise significant market growth through added market development and penetration.

9 Smart Card Human Factors

Many considerations enter into a successful market entry of a technology with new capabilities. Marketing, development, service, and distribution/delivery all contribute to achieving a high level of customer acceptance. However, long-term acceptance and usage growth of the Smart Card will be highly dependent on its *human factors design*. These factors relate to how the customer interfaces with the Smart Card and its accepting work station during all phases of Smart Card use. To a large extent there is a significant period of human factors training and conditioning underway now. That preparation results from the broad current use of FTCs and ATMs.

Which human factors contribute to successful acceptance and use? What does the user want and need to achieve success? What are the problem areas and how are they handled? What are the attributes of a good human factors solution? To answer these questions we need to start with the human factors considerations for the conventional FTC of today. Then a projection of those factors will be made forward to the Smart Card.

EARLY RESULTS WITH CARD-BASED MEDIA

About twenty years ago several early card-based approaches were being developed. They were intended for use in industries dealing with massive use issues, like airlines ticketing and boarding, mass transit ticket sales, and access control and cash dispensers for banking use. There were also noncard-based experiences with telephone direct distance dialing, automobile automatic transmissions, and the controls for new television and stereo units.

Early card use results were informative. There was early public acceptance of the increased convenience offered in the various approaches, as customers were quick to forget the old ways. For example, they left their checkbooks home when starting to use cash dispensers. In fact, they would use self-service units even when agents and tellers were available. Why? Because the self-service units did not hassle customers for identification. The customer could avoid having to write a check. The customer did not have to carry on a conversation with the clerk. The self-service transactions were quicker. The line moved faster and with more predictability. However, the customers also learned quickly that if the self-service units weren't working they had no checkbook available as a back-up. It was at home.

There were some specific human factor results that became known as success factors.

A learning curve of one: The operation of a unit had to be so simple and straightforward that a customer watching another person performing or being tutored through the operation once would allow them to successfully repeat the operation without any assistance.

Avoid computerese: Computer-based product and service developers are so immersed in these terms that it is difficult for them to recognize the lack of public comprehension. Examples of these words include *Enter, Proceed,* and *Key*.

Reduce the user's need to read to a minimum: Users dislike reading, especially if it is very repetitive. Structure the interaction, if possible, to avoid having the users' actions paced by their need to read a display.

Allow an experienced user to increase their speed of interaction: Users with increasing experience in a process want the execution pace to speed up as their experience level develops. Changing built-in delay factors would provide this result. Avoiding an operational dependency on reading a display is an important improvement.

These basic success factors have improved human factors results with card-based self-service devices such as ATMs. However, before settling on the best human factors approach for Smart Cards, we need to examine each of the Smart Card "players."

HUMAN FACTORS IN THE SMART CARD SYSTEM

A number of human factors will influence Smart Card acceptance and usage. Each has a view, and each view needs to be considered and to have an appropriate response.

The user
The acceptor: A merchant for purchase application, a doctor for medical application
The issuer and its marketing department
The issuer's audit and security department
The Smart Card designers, suppliers, and vendors

THE USER

The factors desired in a financial transaction Smart Card by the user includes the following elements:

Convenience: Location, time of use, and procedure.
Ease of getting started: Application, form filling, Smart Card issue process, delay and delivery, and the PIN-based personalization process. Does the personalization process give the user a feeling of security and comfort? Is the process timely and convenient?
Speed of individual transactions: The time factors are reasonably predictable and transactions speed up with experience.
Security: The card is usable in reasonable locations at any time of the day. The user knows his lost or misused or stolen Smart Card liabilities. He should understand the importance and value of the security procedures and suggestions.
Ease of coping with problems: This includes learning, errors and restart, lost or stolen cards, replacing PINs, rules for access, and inquiries and alternative actions if things aren't working.
Viewable interfaces: Card graphics, display readability, simplistic keyboards or entry devices, and illustrated procedures with a left-to-right pattern when possible.

These factors are easily demonstrated in a mass transit system, ATM, or even a long-distance telephone dialing environment. The Smart Card lends itself to these considerations as a result of its evolution from the FTC. Hence, the simplicity of Smart Card use will be considerably easier than a personal computer with its language, procedural, and interactive complexities.

THE ACCEPTOR

In some ways, the acceptor has a more difficult situation than the user. The user obviously wants to make the Smart Card process work for his convenience. The acceptor usually has poorly trained personnel. In fact, the acceptor's personnel become more indifferent when an automatic process is involved. In addition, some of the old standbys will disappear. In banking, the printed hot card list, the authorization call, and the imprinted forms set are all out. What human factors objectives are desired by the acceptor?

Ease of training personnel: A particular need is in handling the old and new solutions during a transition period.

Simplicity in procedures and exception processes: In particular, communicating situations and conditions requiring added handling.

Easy coping with problems: Errors and restart, exceptions or rejected transactions, and workable alternative procedures if the normal solution is not operating.

Security: Ease of detecting and responding to security situations, simple and safe responses to lost, stolen, and unpaid account cards.

The introductory and transition period for Smart Cards will need to recognize and respond to these new characteristics. There will be no substitute for a careful consideration of the human factors in the responses to the new needs.

THE SMART CARD ISSUER AND ITS MARKETING DEPARTMENT

The issuer sees the Smart Card as a new and positive marketing device. Although expensive, the Smart Card's new functions offers enlarged market usage, reduced handling expenses, and a further migration to self-service by users. However, to achieve these results requires a number of human factors considerations and objectives:

Easily understood and usable banking product.

Quick user and acceptor education and participation.

Simple procedures, processes, exception handling, and responses.

Fast growth in market acceptance and rate of use to offset high entry expenses.

Smart Card and interface products work with high availability and throughput transaction rates.

Evolutionary change from current conventional card solutions to the new Smart Card solutions.

These needs are helped by the same basic set of human factor success elements discussed earlier. The issuer faces a challenge in not being able to rely on others to solve his needs. Vendors and suppliers can't provide the market and customer personalization needed for Smart Card market success. The marketing function is faced with naming, introducing, and supporting their new Smart Card offering. Clear identification, early designation of usage points, good initial customer and employees involvement (both issuer and acquirer), and motivation factors to support this effort are musts.

THE SMART CARD AUDIT AND SECURITY FUNCTION

Every new market medium is greeted with early, innovative fraud. In addition, it is greeted by untrained personnel, poorly understood procedures, incomplete response mechanisms, and an unprepared security/audit function. The financial transaction Smart Card offers important transition human factors. Namely, it is a conventional FTC to which Smart Card properties have been added. Therefore, all of the human factors designed into the FTC should be retained as a starting point. The additional Smart Card function and human factors should be added to the existing base during a transition. What are these added factors?

Clear identification of the financial transaction Smart Card and the ability to distinguish it from a conventional financial transaction card.

Procedures to handle both card types, including responses to negative situations.

Clear back-up procedures in case of questions or failure of a basic approach.

Motivational rewards to identify and respond to financial transaction Smart Card negative situations.

Identification of unusual conditions (e.g., an excessive use), simple communication of the condition to the acceptor, development of a quick but safe response, and careful attention to not disturbing or unduly alarming the fraudulent Smart Card presenter.

Security is a two-party war game, to be discussed in the *Security* chapter (page 122). However, a carefully prepared set of human factors considerations applied to the basic procedures and processes should not be considered a permanent solution. Rather, changes will need to be recognized and made as the war game opponents make their moves. The human factors must help to accommodate change in this area.

DESIGNERS, VENDORS, AND SUPPLIERS

Designing and providing the Smart Card and its interface work station has a number of human factors challenges, which cover a rather broad array of considerations.

Design Objectives

Keep it simple: structure, process, and procedures
Ease of use: communications, exceptions, and backup
Ease of learning: offers speed improvement with experience

Electronics

Safety
A 7-day, 24-hour operation
Immediate recovery
Display readable in all locations but especially in the presence of sunlight
Built-in, transparent security functions
Ease of testing that Smart Card is working properly

Mechanics

Carefully designed buttons, controls, and lettering
Left-to-right operation
Clear, unambiguous Smart Card and media interface for insertion and removal
Ease of testing that unit is operating properly

Media design covers a number of important issues, including graphics, colors, labeling and wording, layout, and directional marks. The designer will need to test and tune these factors in order to be assured that the results will produce the right human factors impact.

AN AVAILABILITY CHALLENGE

Smart Cards will eventually be issued in tens of millions. Using locations will be almost anywhere. Today's FTC is primarily a passive, mechanical device. Is it working properly? That is usually easy to determine. Try another card that usually works. Is the card in one physical piece? Is the imprinter forms set readable? Does the ATM accept the card as an initiating device? All these clues tell the card holder that his conventional FTC is probably in good shape and is workable. Few cards fail in their limited life of up to two years.

What will be the equivalent process for the financial transaction Smart Card? All of the conventional tests apply to the financial transaction Smart Card. Physical, visible, and readability tests apply equally well. But what about the electrical chip properties? How does the Smart Card user get a quick and comfortable feeling about that aspect of the Smart Card?

1. Use a built-in diagnostic or self-checking facility that reports successful automatic testing with each use. The result could be indicated on a printer receipt, in a display message, or through a work station indicator light.

2. Provide a self-testing function as part of an easily implemented user function. For example, build the self test into a Smart Card self-service journal inquiry facility. This might be in a bank lobby unit, a home or remote videotex unit, or in a personal computer.

3. Build a Smart Card test function into work stations. This avoids added Smart Card complexity and cost. Remember that there are many more cards than there are work stations. Hence, putting the Smart Card function in the work station or into the job Smart Card will further reduce the user card expense. The correct operation can be indicated after a successful transaction completion.

There is no substitute for a quick and easy way to establish that the Smart Card is working properly. One word of caution. The financial transaction Smart Card does not open all storage areas. Other storage areas are available only with the insertion of the proper PIN value for the card. Nothing in the quick tests must countermand any of those security controls. Nor should they give the test equipment access to areas in a manner that defeats the security controls. The quick tests might be PIN-value actuated but not involve external access to the sensitive areas. A quick indicator of proper operation will help to maintain user, issuer, and acceptor confidence and support in the use of Smart Cards.

EXCEPTION, ERROR, AND FAILURE RECOVERY

Something will eventually go wrong in any system. What is the recovery path? This may be a particular problem with the great circuit density of the modern integrated circuit chip. The routine chip failure rates may be so low as to provide a long period between failures. That means that no one will remember what to do when a failure does occur. Hence, an easy-to-follow, uniform process is needed. For example, provide a test user card with every work station. If there is any doubt about the customer's Smart Card, remove it and substitute the test card. Put the test card in and let it display a standard response. The same approach can be used with the job or device cards. It is very important that an easy, quick, simple, and reliable indication be available for these infrequent events.

SUMMARY

A leading banker once told me that customers would do anything with self-service devices that would ensure their successful operation. They would lean out of the window of a panel truck to reach the low-level controls of a drive-up automatic teller machine. They would, in short, tolerate a lot of product preparation mistakes. Unfortunately there is another side to this willingness to be tolerant when a service is highly desirable. That is, the customers quickly lose patience with units or services that are not working properly. They will vandalize. They will stop using the service. They will withdraw accounts to go to a more available unit at another CONVENIENT institution.

Putting all of these factors together suggest the following human factors objectives for Smart Cards:

Achieve a learning curve of one.
Keep it simple.
No computerese.
Reduce the reading.

Let the experienced user speed up.
Make the Smart Card work station operating times reasonably predictable and consistent.
Provide a simple indication that something is not working and an easy way to cope with that problem.
Maintain high availability, including a simple and direct way to check that a Smart Card is working properly.
Clearly indicate which Smart Card is acceptable to a work station.
Keep all procedures simple in order to insure that a rare failure of any Smart Card (user, job, or device) is consistently handled.
Consider providing institutional rewards to Smart Card acquirers for detecting and correcting Smart Card problems quickly.

Successful human factor designs are produced by well-trained experts. They need to be provided with a reasonable amount of money and time in order to do their job properly. This chapter is designed to give you some insights into these needs in order to understand and help the professionals to do the right job.

10 Smart Card Survivability

Survivability is a simple concept. A Smart Card, when produced in an economic manner, must be capable of correctly performing its expected functions for the duration of its intended life. In the process of achieving that objective, the Smart Card will need to withstand the whole spectrum of abuse, misuse, and mischief that issuers, users, and acquirers are capable of perpetrating. The challenges will be great. Remember the Unofficial and Unhelpful Role of FTCs described in Chapter 3, page 31. Those demands apply equally as strongly to the Smart Cards.

SURVIVABILITY

Survivability is a concern at each step in the Smart Card life cycle (see Card Life Cycle on page 23). It starts in the design and manufacturing phase.

Physical Package Design

The thickness and stiffness characteristics of the Smart Card physical package provides structural support to the enclosed chip(s). Obviously, more thickness and stiffness is better. However, the new design may also need to be physically compatible with a prior design.

The physical compatibility for both old and new card types may be needed to provide several years of transition from pre- to post-Smart Card devices like ATMs.

There is no easy answer to this requirement. Several approaches have been used to minimize the impact of thinness. One approach is to locate the chip at a point of least stress. Another is to use a thinner chip, remove some of the stress, (e.g., not embossing the card).

Only recently have financial transaction Smart Card packages achieved the thinness prescribed by international card standards. One approach has been to put the chip into a carrier. The carrier, in turn, is put into the card. The carrier is designed to protect against heat, contamination, and provide contacts. There are several carrier insertion techniques, as there are several chip mounting techniques for the carrier, including the use of wire leads or the use of tabs with film mountings for high-volume insertions.

Chip Protection

The chip needs to be protected from a variety of electrical, mechanical, and chemical dangers.

1. Chemical: There may be vapors from plasticizers in the package, and petroleum and water in the environment. The draft standard for the financial transaction Smart Card does not meet this challenge. It does not require the chip to be hermetically sealed. Note that *hermetical* sealing is a much more demanding specification than sealing.
2. Mechanical: These relate to the pounding, bending, and flexing of the physical package. Several possible protections are possible, such as an inner chip package for added strength, added stiffness in critical locations, and the eventual elimination of embossing.
3. Electrical: These are perhaps the most challenging because they are not the most obvious. Heat dissipation is probably more serious in the manufacturing phase. The added heat results from the extended power-on time required to initially load the chip. If not dissipated, the added heat could result in chip burn out. The heat could also, at a lower temperature than chip burnout, distort or weaken the plastic package. A maximum heat dissipation level and surface temperature of 120°F is specified in the

draft standard for the financial transaction Smart Card. If the chip is enclosed in an inner package, then that package may need to provide a heat dissipation surface. This surface may need to be matched with a cooling surface or facility in the work station interface device.

4. Chips may also be susceptible to the effects of static electricity. The effects may be sufficient to destroy circuits or to distort the electrical operation of the chip. Two alternatives might be considered. One is the use of a conductive coating to remove the static charges upon insertion into a work station. The other is to design the inner package to provide electrostatic protection.

Contacts

Contacts are the electronic conduit to the chips, and their protection from the world must be considered.

1. Contamination: The surface of the electrical contacts is exposed to several substances increasing the resistance to the current flow. One type of contaminant might be dried glue, originating from the mailer package to which the card is attached for mailing purposes. Other contaminants might be from the signature panels of other cards together in a tight wallet, or chemical contaminants that oxidize or corrode the contact surface. Generally, the solution is a contaminant resistant surface plus a contact interface with a wiping or probing action, intended to overcome the surface contamination.
2. Contact assignment: This problem relates to which contact has the high-voltage assignment. Assume the assignment is to a contact furthest away from the card edge. The card acceptor may not be working. It leaves the high voltage turned on even after the card has been withdrawn from the unit. As the next card is moved into position, a wiping action places the high voltage on each of the intermediate contacts, possibly damaging or destroying the other circuits. In fact, the wiping action may have occurred on the removal of the previous card and damaged it. That damage will not be detected until the card is used at another time and at another work station.

There are several solutions possible. One is to retract the contact mechanism on card insertion or removal. Another is to design the other circuits to withstand an overload until the card is properly seated, or design an electrical interlock that is activated by card insertion. I doubt that this issue will be resolved at the standards level, since when I proposed a change in assignments in the draft standard it was rejected. Why? Some vendors argued that their terminals and work stations always operated properly. Hence, this was no problem! Unfortunately, there are vendors that don't build safeguards into their work stations. That is the problem!

ISSUING SMART CARDS

The issuing process exposes Smart Cards to more pounding, contaminants, and mishandling. For example, after the manufacturing process it will be necessary to personalize the content of each Smart Card to its user information. The financial transaction Smart Card will be personalized by embossing the account number, by recording the stripe content, and by loading information into the chip(s). This loading includes account information, security and service controls, and the PIN and bank access codes. As discussed earlier, the embossing process applies serious and abrupt forces. The contacts, and hence the chip, cannot be far away from the embossing area. One alternative might be to emboss first and then insert the chip. This has been discussed, although I do not know of a Smart Card supplier that is actually producing cards in this manner.

Following the personalization process, the Smart Card needs to be delivered to the user. In many cases this will be done by mailing. This is another set of machines, carriers, envelope handling and sealing, sorting, routing, and physical handling. The Smart Card designer must anticipate this handling and its range of forces, stresses, bends, and twists. Equally important is the need for a simple but effective test that advises the Smart Card receiver that the card is working properly. A simple journal inquiry facility might suffice for this need.

Another alternative is to bypass the delivery process used with today's FTCs. For example, a batch of cards are delivered in bulk to a location at which the Smart Card user can have the card personalized and picked up. The personalization process would be completed by the user entering a PIN value into the personalization equipment. The procedure would solve several concerns simultaneously.

SURVIVABILITY IN USE

Several dangers to the Smart Card exist in its daily use. In general, user education is a weak solution to most problems pertaining to cards. Rather, the designer and the issuer will need to anticipate the problem in the design and implementation.

Environmental

A wide range of temperatures and humidities will be experienced. From the depths of winter in the North to the heights of summer in the South are a reasonable range of concern. These environmental conditions are a problem primarily for the plastic package.

Physical

Imprinting with an embosser is probably the sharpest physical impact that the Smart Card will need to absorb. However, casual events such as dropping, bending, jamming, and tapping cards on hard surfaces are not uncommon events.

Expected Card Life

The draft standard for the financial transaction Smart Card indicates an expected life of 36 months. With 10 uses per month, that produces almost 400 uses for the card over its intended life. That is a lot of opportunities for use, abuse, and misuse. *Let the designer beware!* Let the user be cautioned. Let the issuer be willing to pay for an "adequate" Smart Card product. Let the chip supplier shift his sights from the protected world of electronic equipment to the world of people-carried, used, and abused microcomputers-on-a-chip.

LINES OF DEFENSE

There are two basic approaches to survivability. One is to have a better understanding of the survivability requirements and specifications. The second is a willingness to invest in survivability design.

Better Requirements and Specifications

A close cooperation between the vendor, supplier, and issuer is important. Agreeing early on common needs is the initial goal. Testing of designs for unanticipated weaknesses and exposures is key. Designing to protect against known hazards is required. All of these steps suggest an early awareness that the Smart Card is quite different from its predecessors. The early conventional FTCs were solid, passive, and based on issuer experience with large volumes of cards. In contrast, the Smart Card is a package containing dynamic elements and is still very early in its concept, life, and field experience. There is a great deal to be learned. Intensive testing will be the foundation for a predictable future.

Invest in Survivability

Many of the solutions discussed will not be cost-free. Hermetic sealing, heat dissipation, static electricity grounding, and other design features are needed. These considerations have been successfully provided in a broad array of consumer, educational, and personal computer chip electronics. These extra costs will be essential in the early design phases. More important, they are needed to build a base of issue creditability with the issuers and the users.

Another form of investment is to use a less challenging chip design. This could be a less dense chip, a smaller chip, or a lower function chip. These are alternatives for entry products that might help early survivability. There will be a sufficient time and experience later to build up the chip size and complexity.

ENTRY PLANS

A number of guildelines from earlier card entry efforts are worth considering:

Expect the unexpected.
Build the shipped volumes gradually.
Track actual usage and card performance.
Examine returned, expired, and collected cards.
Set a shorter planned field life for entry cards (e.g., 6 to 18 months).
Extend expected field life gradually as the shorter period successes are confirmed.

Encourage customer and acceptor reports of unusual experiences or field results.

Test in-use cards on a sample basis. Look for signal degradation, physical package deterioration, contact erosion, and contact contamination.

Collect work station use statistics on Smart Card interface errors and work station malfunction.

These are really common-sense guides. An early investment in tracking, solving, and confirming Smart Card survivability is essential. In the long run, creditability for the Smart Card will be based on specific experience. An effective marketing effort will get customers to ask for and use the Smart Cards. Only successful and sustained performance will be able to maintain user and issuer interest. Actual performance is the key to achieving a high level of Smart Card usage—beyond the short interval of user and market curiosity.

11 Smart Card Security

Security is like beauty. Its importance, its proper solution, its cost acceptability, and its marketability impact are values in the eyes of the beholder. Current financial transaction card security architecture has some shortcomings, but the providers seem willing to accept them. In fact, it is only recently that increased card fraud has started to demonstrate some of the dangers. There are other frauds yet to be experienced. Some of these are expected with the magnetic-stripe card.

Frequently United States bankers say things about on-line controls. Yet on-line frauds are beginning, for example, line tapping and line interventions, that provide false authorization responses. These false responses are mistakenly accepted at a point-of-sale as a legitimate authorization response.

The purpose of this chapter is not to be primer on card-based crime and fraud. Nor is the content intended to provide easy solutions with guaranteed results. Those answers are left to the entrepreneurs and to the industry "experts." They will have to negotiate the compromise solutions. They will have to balance cost and investment against reasonable results and protection assurance. They will have to educate and enlist the help of card users and acceptors.

The objectives of this chapter are, like a very good security solution, very simple:

Describe the problems, particularly the new ones that will appear with the Smart Cards.
Identify the lines of defense, including the new responses available with the new product and application functions of the Smart Cards.
Extend the opportunity matrix to include the security functions.

The most challenging aspect of security is a basic set of experiences with cards:

There will be early, clever, and unexpected security attacks.
We don't know what we don't know—therefore, security responses must be designed to expect the unknown.
There is no such thing as a secret. In fact, the best defense is one that gives reasonable protection when the protection details are known to the attackers.
Any security architecture must be treated as a two-party war game. A reasonable investment is made in a security plan and its implementation. The opponent occasionally makes an unexpected countermove. The architecture must then be evolved to respond with the next reasonable level of defense.

In most cases, the Smart Cards are one part of a total solution system. For example, the financial transaction Smart Card is one element in an electronic banking system. As such, there are a number of security elements that apply to the total system. Although we are not planning a discourse on the broad electronic banking security subject, we can state a basic set of pertinent issues that reflect the other issues:

Bank Protection Act (National): Requires designation of a security officer and various forms of mandatory actions and reports.
Regulatory requirements: Requires specific actions in support of regulatory rules and legislative acts (e.g., regulation P of the Comptroller of the Currency specifies protection requirements for cash dispensers).
Auditors and examiners: Requires appropriate controls, reports, and aggressive probing in all electronic system aspects.
Federal Deposit Insurance Corporation (FDIC): Participates in bank examination and status but also offers support in security needs and responses.
Underwriters Laboratories: Requires vendors to be listed as comply-

ing with UL specified test criteria (e.g., UL criteria for safes and automatic teller machines).
National card authorization and interchange systems.
Card suppliers with secure manufacturing and delivery facilities.
Issuer-based card and PIN issue systems.
National Financial Services Standards (e.g., ANSC X9 standards for PIN Security (9.8, 1982), and Message Protection (9.9, 1982).
National card issuers security organization.
National and state privacy and security legislation.
Bank industry card fraud task force.

Suffice it to say, total system security is an issue with which the Smart Card needs to integrate its efforts, architecture, and responses.

SMART CARD PROTECTION

To assist in this description we will focus on the three fundamental Smart Card configurations defined previously (see page 74). These are the: (1) increment, (2) interchange, and (3) interactive smart cards. The other important dimension in a security discussion are the levels of attacker sophistication. Let us consider the following three levels of card fraud skill:

1. Casual: The average card holder, bank employee, or merchant employee.
2. Determined: The very capable amateur or student, the experienced colluding merchant, the street-smart hot card specialist, or the knowledgeable issuer insider.
3. Professional: Long experience combined with a willingness to make an investment and a willingness to buy or acquire know-how.

SMART CARD SECURITY

There are a variety of Smart Card security problems that need to be faced as basic issues.

1. Information flow between the Smart Card and its interface device: Any information moving to or from the Smart Card is directly tappable and copyable for all Smart Card types. The attackers

are from levels 2 and 3. Also, external signals may be introduced to provide false responses to the Smart Card.

2. Simulated Smart Card: The information flow to and from the interface device from a Smart Card is simulatable. That is, the interface device can not distinguish between a legitimate Smart Card and a simulation version. The simulated Smart Card accepts and responds to all signals in the prespecified manner. This is true for all attacker levels and for all Smart Card types. The casual attacker will probably buy a simulated Smart Card from a higher level of sophistication attacker. The simulated card need not be the same shape as a Smart Card. A microprocessor on an extended size "bread-board" base might provide the appropriate signals to the electrical contacts while providing physical space for the simulation equipment.
3. Smart Card replacement during the transaction: There may be a time "window" within which an authorizing Smart Card can be removed before the responding signals attempt to write into the Smart Card. In that way the response signals are written into another card whose lower balance is not used to start a transaction. This attack may be available for all types of Smart Cards. It is implementable by all attacker levels.
4. Providing incorrect control information to a Smart Card: The smart card requires today's date. The date is used by the Smart Card logic to decide if the available balance fields should be updated to the maximum authorized balance. If the control date is made available on a premature basis, then the available balance may be artificially high. Given the ability to make several date changes, the available balance might be reusable with every date reentry. This form of data attack is usable for each Smart Card type. It is implementable for all attack levels.
5. Merchant or acquirer frauds: Any action or procedure by a clerk or merchant will be subject to mistakes, errors, deliberate attacks, and nonprocedural actions. This is an exposure for all Smart Card types and for all attacker levels. For example, the clerk enters two transactions while the customer believes that only one transaction is occurring.
6. Misappropriate the Smart Cards and/or work stations: All Smart Cards and work stations may be stolen, borrowed, disassembled, modified, used for attack testing, or otherwise interfered with independent of any one knowing. This is a danger for all Smart Card types and all attacker levels.

7. Attacking the Smart Card personalization process: Any portion of the Smart Card issue, personalization, initial delivery processes, or security protection response in which a Smart Card issuer employee is involved. This applies to any Smart Card type and may be from any attacker level.
8. Relying on the Smart Card user: Any security protection or response scheme that relies on the user must be suspect.
9. Security needs: In applications using financial transaction Smart Cards, there is a basic set of security functions required of the Smart Card. These are:

   Valid card
   Valid user
   Protected controlling data and limits
   Usage record
   Transaction authorization
   Information flow protection

Not all of these needs have perfected Smart Card responses. Responses will be proposed. Experience will test and refine the responses. These in turn will be tested further in the continuing War Game.

This list is not meant to induce a panic. Nor is it intended to be an exposé of security holes. Rather, it is intended to portray the outer range of potential exposures. A similar list could be constructed for today's FTC. However, as the old Chinese proverb proclaims—To know that you have a problem is half of the solution.

This list reminds me of a situation in Japan over a decade ago. The Japanese bank card issuers decided it would simplify their system security if they put each individual's PIN on the magnetic stripe of the individual's bank card. I was sent to Japan to demonstrate that the unscrambled PIN on the magnetic stripe was readable by almost anyone. I brought with me a bottle of magnetic recording developer, which can be bought in a corner electronics store. It is used by magnetic recording service personnel to test that a recording process is working properly.

I met with several members of the technical committee. I borrowed a striped card, sprayed the stripe, and showed the members the PIN recording and read the value to them. They were impressed! They had not seen that demonstration before. The committee asked to be excused to discuss the subject, then returned a few minutes later. The chairman thanked me for my presentation. He then advised me that the committee didn't really think that they had a problem, noting that the banks didn't usually issue cards to Americans!

Actually the losses incurred as a result of this vulnerability have apparently not been great. However, I'm pleased to report that now, after some ten years of large-scale issue, the PIN values are being removed from the Japanese bank cards. Similarly, the exposure list above doesn't mean that all of the exposures will necessarily be experienced.

LINES OF DEFENSE

Any security response requires a basic plan of action. There are guidelines that have been proven useful in prior implementations.

1. Vital data: All vital data requires either physical protection or information protection. Vital data includes decision results (e.g., a yes or no authorization). Physical protection means dual access control or automatic destruction of vital data when an unauthorized physical access is attempted. Information protection means that the information is encrypted, scrambled, or combined with other unknown data so as to hide its content. (*Hide* means that the amount of work required to determine the vital data is in excess of the value of that hidden data). One form of physical protection might to be fully enclose the Smart Card while it is actually in use.
2. Security processing area. This protected area of the electronic system is required in both the Smart Cards and in the interfacing work station. This area contains both logic and storage that are not externally readable. Any attempt to enter or read the content will automatically result in destruction of the vital data. Data may be entered into the protected area. Data may be removed, but only in a protected form.
3. Assignment of responsibilities: Any security plan requires a clear assignment of responsibilities:

Card supplier: Provides the required security function in the Smart Card and a secure manufacturing and physical distribution capability.

Work station vendor: Provides the required security function and supporting software.

Financial institution or Smart Card issuer: Generates, issues, and delivers vital data in a protected manner.

System provider: Treats vital data in a secure and protected manner at all times.

Smart Card user: Receives and protects vital data (e.g., his own PIN number).

Smart Card acceptor: Protects the work station and the job Smart Card at all times from casual borrowing and exploration, and removes his job card to a separate area when it is not in use (if the job card is removable).

The application lines of defense are also needed to provide full system security.

All transactions must be journalized in at least two separate journals.

All transactions will have vital fields subject to application controls, e.g., maximum amounts.

All transactions are subject to sensitivity tests such as frequency of card use or total value of use in a prespecified period.

All exception conditions requiring on-line authorization are defined and planned responses are set.

Hot lists are not the opposite of positive card lists. Positive lists must also be maintained to include control of unissued account numbers. In most situations the number of unissued account numbers is many times the size of those issued. A possible solution might be for all account numbers to require a check digit that can not be determined by an easy technique. This would detect attack techniques attempting to guess at unissued account numbers.

EXAMPLES OF DEFENSE

The incremental, interchange, and interactive Smart Cards were used previously to demonstrate the range of security exposure. The general elements of security for each Smart Card type is shown in Figure 11.1.

Each of the card and work station types fit with an interface device for its operation. This combination needs to observe some security lines of defense:

The interface for the user's Smart Card or work station and the vital data entry device (e.g., a PIN pad) should be in the same physical unit, not separated by a cable connection.

The vital data entry process should not be visible to the acquirer personnel or anyone else.

The vital data control (e.g., a PIN) must be re-entered for each transaction to prevent multiple transaction entry without Smart Card user knowledge.

	SECURITY LINES OF DEFENSE		
SMART CARD TYPE	CARD	APPLICATION	SYSTEM
INCREMENTAL CARD	IDENTIFICATION CODES	INCREMENTS DESTROYED TO PREVENT REUSE	OPTIONAL JOB CARD CONTROL CODES
INTERCHANGE CARD	PLUS ACCESS SECURITY	VITAL DATA PLUS APPLICATION CONTROLS	ON-LINE CONTROL PLUS JOB CARD HOT LIST
INTERACTIVE WORK STATION	PLUS SESSION LOGIC CODES	LIMIT AND DATA BASE ACCESS CODES	ALL ON-LINE CONTROLS AND SESSIONS

Figure 11.1.

Removal of the user or the job Smart Cards from the work station must immediately stop the transaction and must destroy any user's vital data.

An exception process using on-line authorization must be provided for high value transactions in any environment using store and forward system attachment mode. One trigger for an on-line authorization should be a remotely settable random or sample mode. The sample rate may also be settable by the job Smart Card content. The work station should not indicate if an on-line authorization is taking place. This means that local authorizations may include a planned response delay so as not to be distinguishable.

Some applications use the job Smart Card for added control by providing a hot list of bad account numbers, but such a list may not have sufficient capacity. Also, as indicated previously, a hot list does not check for unissued account numbers. In this case, several rules should be observed.

Any cards not issued locally must have their transaction authorized on-line, or the transaction must be rejected if it is over a preset floor limit.

All local cards should have a control code in the secret area that indicates that it has been on-line authorized at least one time during a preceding period (e.g., 30 days). If not, the transaction must be on-line authorized or the transaction must be rejected if it is over the floor limit.

Any card that contains control fields exceeding a specified usage frequency or allowable usage value or in which the account number does not have a proper check digit, must be on-line authorized.

FUTURE SECURITY

There are a number of new security solutions on the horizon, in response to market demands. In most cases the new solutions will be equally useful to conventional cards and on-line systems. However, they are going to enhance Smart Card function and capabilities. Here are some of the coming changes as I see them.

1. Personal identification verification: A number of new techniques are in development. Signature verification by signature dynamics, identification by retinal scanning, and verification by hand geometry are three. These techniques and others will be tested and tried. A winning technique will eventually emerge after several years of trial and use. The PIN value will not disappear. Rather, it will be used in existing environments and will be useful for many years as a backup to the emerging technique.
2. Secure card properties: The financial transaction Smart Card has three major components. These are the plastic, the magnetic stripe, and the chip. These are three separate components. The

stripe content could, for example, be easily replaced with a content whose identification and control values were different from the other two components. Yet, in a device that reads only the magnetic stripe, the identification in the embossed number might be totally different. Thus, a backup authorization process done manually with the plastic would produce a different result from using the magnetic stripe. There is a need to tie these three elements together. This feature is generally referred to as a *secure card property*. One technique proposes a nonerasable stripe with a unique identification code that is physically fused into the card surface. Thus far these techniques have not had great acceptance, because 85 percent of the industry problem is the bad debt of the person to whom the card was issued and not fraud. But this is starting to change. It is not clear what role the stripe will play with the Smart Card, however, stripe-to-chip security will be necessary for probably a decade of transition to the financial transaction Smart Card.

3. Private and public key encryption algorithm: The national encryption standard—the Data Encryption Algorithm (DEA)—uses the same key value to encrypt and to decrypt a piece of data. This private key system creates an implementation need, namely, how to distribute and protect the key values. Since each pair of communicating parties must have an independent key, it becomes a major burden of the key management system to properly administer. A new type of algorithm called *public key* is now in development. Each individual in the system is assigned a unique pair of key values, one for encryption and the other for decryption. This simplifies the key management requirement. The holder of public key-based encryption can not "see" previously encrypted data from other pairs or the same encryption value. Thus far these techniques are much slower than DEA. The required key values are very long and have not been validated by a broad group of experts. This validation would assure users that there is not a simple solution. Also, the validation would establish that the work required for a brute force solution is significantly large. However, the feature of the one-way key is very attractive. A good deal of effort will be spent to successfully develop this type of encryption algorithm. When perfected, it will co-exist with DEA for a long time. It is not a matter of replacement, but the use of each technique in its area of strength.
4. Key management: Smart Cards may be designed to have restricted or secret access areas. These areas may be used as a *key manage-*

ment device. This area can store a key value for DEA, public, or other key-driven algorithm use. This restricted access area prevents unauthorized access to or inspection of the key value. However, the value can be used internally for algorithmic calculation. This portable key management can significantly simplify the key management process. The current (mid-1985) Smart Cards do not have sufficient capacity to execute DEA. Hence, controlled release to an external DEA device is possible until adequate capacity becomes available.

CURRENT SOLUTIONS

The factors and guidelines discussed above are important considerations in the most vulnerable Smart Card security area, namely, the interchange Smart Card being used in a store and forward or remote mode. The current French tests use other, complex control procedures. The procedures use proprietary encryption algorithms with a multiple set of controlling keys. One key controls the job card data movement to the acquirer's bank, another key controls the journal data from the user's card to the issuing bank. Most of the details for these procedures have not been described in detail. Nor is there any evidence, test results, or independent assessment that they work.

The French Telepaiement (PTT) system offers one security approach. See Figure 11.2, provided by Dr. S. Weinstein from his *IEEE* article of February 1984. It shows several Smart Card security challenges. The suggested steps in the diagram are self-explanatory. They start with PIN entry (1), generation and transmission of an authentication code (2,4,5,6), a comparison of the code by the controlling computer at the bank and the return of an approval signal to the remote terminal (8).

This process has several significant problems:

1. The PIN value moves in the clear from the remote terminal to the card. The contact signals may be monitored and copied.
2. There is no such thing as a secret algorithm. It is known to manufacturing, development, and service personnel. The algorithm may be simulated, its output captured for given account numbers, or previously captured values sent to the computer.
3. The computer comparisons at the bank may be accessed or compromised and an incorrect decision returned to the remote terminal.
4. The return signal is as valuable as the original request and supporting PIN value, and so needs protection.
5. The process is clearly on-line and does not portray the proposed solution for off-line or store and forward operation.

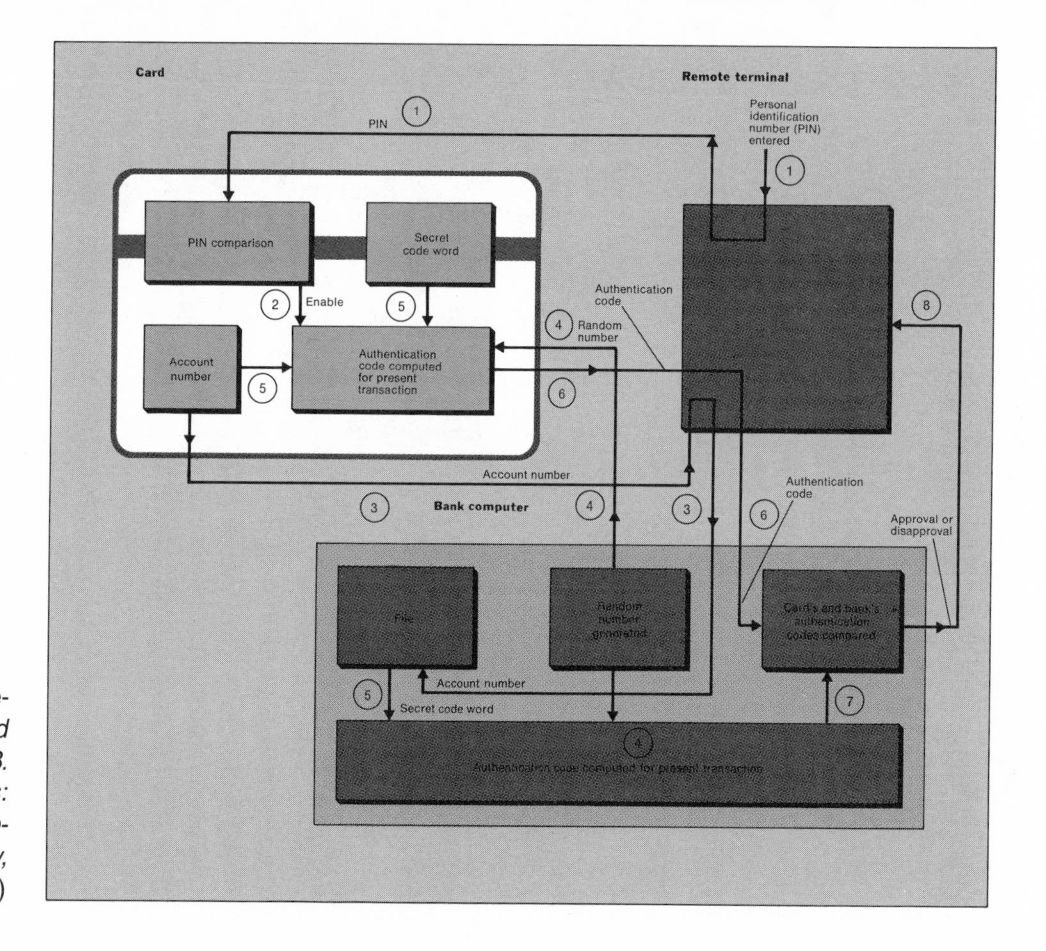

Figure 11.2. The French Telepaiement system. (*Reproduced with permission from Steven B. Weinstein: "Smart Credit Cards: The Answer to Cashless Shopping,"* IEEE Spectrum, *February, 1984. Copyright © 1984, IEEE.*)

These potential security problems in a proposed solution demonstrate the challenges to the security architect and designer. In a magnetic-stripe card solution, the entered PIN value would be encrypted and sent to the controlling computer. A unique variable would be included in the encrypted PIN field. Also, the return response would include a unique variable in each message that the remote terminal could validate before taking final action.

CERTIFICATORS

The Smart Card introduces a new level of Financial Transaction Card security. Included is a PIN-based access to the card content for transactions. Also, the in-card logic allows an algorithmic computation of a unique "signature" for each transaction. This establishes that an "authentic" card has been presented. Not all transaction points have sufficient transaction volumes to justify using a terminal to take advantage of the new Smart Card security features.

The Certificator is a low-priced (e.g., $50), Smart Card device that establishes card authenticity, validates the customer-provided PIN, and provides a certification code whose unique value includes the keyed-in amount of the transaction. The Certificator consists of a Smart Card connector slot, a PIN pad, a 6-digit display (LCD), and an internal battery (with a battery charger connection) (Figure 11.3).

The operational sequence is: (1) Insert the customer's Smart Card. (2) Authenticate the Smart Card. (3) Customer keys PIN and the display confirms that the correct code was entered for that card. (4) The transaction amount is keyed. (5) A computed electronic signature is displayed and copied to the transaction form set by the merchant. (6) The customer's Smart Card is removed.

Certificators are intended to be an economic response to a significant investment in Smart Cards. They are a practical way to control losses at nonterminal merchant locations. The Certificator is also useful in wholesale banking. A test code is received from a remote location. It is entered as a "transaction" amount in a certificator with a user Smart Card and entered PIN. The combination produces a unique output value that is then returned to the remote control point for user and transaction validation.

SECURITY STANDARDS

An international standards security effort is intended to produce a plan of action for financial services industry Smart Card security. The ef-

Figure 11.3. An off-line certificator *(Courtesy of Microcard Technology, Dallas, Texas.)*

fort is focused in an International Standards Organization working group (ISO 68/8/7). Its end product will initially consist of three security modules, as follows: (1) Card life cycle, (2) transaction process, and (3) key management.

These standards are the minimum common steps taken to assure all participants of a reasonably secure process. However, they allow individual card issuers to go further, for example, adding a personal identification verification function to their Smart Card usage.

SUMMARY

It is our job in evolving the Smart Card security system to maintain the following:

A reasonable level of security.
A reasonable set of sensitivity tests for the unknown.
A reasonable set of back-up actions.
A reasonably understandable and implementable security plan.
A reasonably serious concern about the most dangerous participants—our employees and our customers.
A reasonably low cost of security that compares favorably with the risk.
A reasonably simple design that is acceptable as an international standard.
A reasonably transparent security process that is more difficult *not* to use.

Back to the start of this chapter. What is the definition of "reasonably"? It is like the definition of beauty. It is in the eyes of the beholder!

12 Smart Card Economics

The economics that will drive the adoption of Smart Cards include the following:

Increased revenue opportunities
Reduced operating expenses
Improved controls with reduced losses
Cost avoidance
Increased productivity
Needed functions not otherwise available

However, an important starting point is the projected cost of Smart Cards. They will be used in large volumes, and any proposition is heavily dependent on the mass-production costs.

In this chapter, economic justification will be considered for the three fundamental types of Smart Cards considered previously (page 74). The major Smart Card attributes will be reviewed, the new Smart Card justification factors identified, and the operational improvement justification examined. Finally, future financial transaction Smart Card prices will be estimated.

PROJECTED CHIP COSTS

Future chip costs are a function of several factors. These include the circuit densities, production volumes, microprocessor word size, chosen

TABLE 12.1

	MICROCOMPUTER-ON-A-CHIP (word size in bits) (costs in dollars)				
	4	8	16	32	64
1977	6.00	11.00	100+	100+	100+
1982	2.00	3.50	10.00	100.00	100+
1987	.90	1.40	3.50	22.00	100+
1992	.60	.85	2.20	12.80	50.00
1997	.45	.70	1.40	9.00	30.00

Source: "Future Information Processing Technology," August 1983, Institute for Computer Sciences and Technology of the National Bureau of Standards. Reprinted in *Datamation*, December 1983.

technologies, and date of production. Table 12.1 is an estimate of chip cost through 1997.

These chip costs are for general purpose devices. Selection of the word size will probably be determined by the required transaction execution time. In most cases the increment Smart Card will not need a microprocessor. However, it might be the least expensive implementation because of the volume prices. The interchange Smart Card will require both storage and logic. However, its performance demands are probably not high, so a small word size, 4 or 8 bits, microprocessor will suffice. The interactive Smart Card will be a full work station. However, it is not clear if the Smart Card functions will be included in the general-purpose microprocessor, or if a separate dedicated chip for Smart Card functions will be appropriate. At this point it is a reasonable guess that the separate Smart Card function chip is the way to go because of its security, control, and dedicated information protocol requirements. Therefore, the Smart Card portion of the interactive Smart Card work station will have the same performance level and capacity chip as the interchange Smart Card.

Storage is another important element of the Smart Card. The cost per bit of storage (RAM type) has been estimated for the same time period as the microprocessor-on-a-chip estimates above. The costs of storage chips are also a function of size, density, technology, access time, and release date. Table 12.2 is an estimate of storage cost per bit through 1997.

These costs are for separate storage chips. In most cases, the Smart Card will use a combined or custom design chip. However, the com-

TABLE 12.2

	STORAGE COST PER BIT (on chips) (in millicents per bit)		
	SLOW SPEED	MEDIUM SPEED	HIGH SPEED
1977	80	200	700
1982	10	20	80
1987	1–2	3.50–4.50	15–25
1992	.1–.5	.5–1.2	3–8
1997	.01–.2	.2–.4	.7–3.5

bined function chips are starting to appear on a general-purpose basis. That is, both logic for a microprocessor and its required storage will be on the same chip. Also, other technologies will be used, such as non-volatile memory. This is required for the interchange Smart Card, which has no internal power source between work station usage. These alternative technologies may also add costs to the chip evolution costs. Hence, any overall cost assessment is an approximation only. However, by combining the projected costs they create some interesting speculations.

Assume an interchange type of Smart Card. Assume that it consists of an 8-bit or 16-bit microprocessor-on-a-chip. To that add 16,000 bits (16K) of slow-speed storage or 32,000 bits (32K) or medium-speed storage. These combinations yield the cost projections of Table 12.3.

These are, at best, very rough cost indications. To them must also be added the expense necessary to create, test, and manufacture custom chip combinations and the expense necessary to package, issue, personalize, and deliver the Smart Card. How does the issue cost of a financial transaction Smart Card compare to that of a conventional FTC? Consider Table 12.4.

TABLE 12.3

	SMART CARD COMBINATIONS					
	LOW CAPACITY			MEDIUM CAPACITY		
	8-BIT MICRO	16K STORAGE	TOTAL COST	16-BIT MICRO	32K STORAGE	TOTAL COST
1977	$11.00	$12.80	$23.80	$100+	$64.00	$100+
1982	3.50	1.60	5.10	10.00	6.40	16.40
1987	1.40	.24	1.64	3.50	1.28	4.78
1992	.85	.05	.90	2.20	.27	2.47
1997	.70	.02	.72	1.40	.10	1.50

TABLE 12.4

FINANCIAL TRANSACTION CARD ISSUE COSTS (in dollars)

	CARD	PERSON-ALIZE	DELIVER	TOTAL COST
1982	.20	.20	.60	1.00
1987	.30	.20	.80	1.30
1992	.40	.20	1.00	1.60
1997	.50	.20	1.20	1.90

The card costs are increased to reflect card technology increased costs for new features such as secure card properties. This projection shows a 30 percent increase in issue costs in 5 years in constant dollars. The FTC is a mature project and the personalization process is well mechanized, meaning that both of the costs will remain rather flat in constant dollars. Delivery costs, however, are tied to postage rates, which are expected to continue increasing, even without an inflation factor.

Using a comparable cost structure, let us consider the financial transaction Smart Card evolution in Table 12.5.

Combining the issuing costs for both the financial transaction card and the financial transaction Smart Card, we get Table 12.6.

These figures suggest that the cost of an issued financial transaction Smart Card will range from 1.5 to 2 times more than a conventional FTC by 1994. In the shorter range (by 1987), the Smart Card costs will be about 3 to 6 times that of a conventional card. A recent an-

TABLE 12.5

FINANCIAL TRANSACTION SMART CARD ISSUE COSTS (in dollars)

	CARD		ISSUE		TOTAL COSTS	
	LOW SPEED	MED. SPEED	PERS-ONAL-IZE & PCKG.	DELIV.	LOW SPEED	MED. SPEED
1982	5.10	16.40	2.00	1.00	8.10	19.40
1987	1.64	4.78	1.25	.80	3.69	6.83
1992	.90	2.47	1.00	1.00	2.90	4.47
1997	.72	1.50	1.00	1.20	2.92	3.70

TABLE 12.6

	DELIVERED CARD COSTS (in dollars)		
	FTC	FT SMART CARD	
		LOW SPEED	MEDIUM SPEED
1982	1.00	8.10	19.40
1987	1.30	3.69	6.83
1992	1.60	2.90	4.47
1997	1.90	2.92	3.70

nouncement said that 1.4 million financial transaction Smart Cards would be issued in France in 1984. The estimated card cost was given as $2.40. An added packaging, personalization, and delivery expense of say $3.00 would put the issued costs at $5.40. That coincides approximately with the cost in Table 12.6 (low-speed column, interpolated to 1984). These cost differences are not large. In fact, the issued cost of the current FTC probably came down by a factor of 2 during the 1970s. Hence it is reasonable to anticipate that the large 1983 cost differences will diminish rapidly. The price changes in hand calculators is an excellent example of this type of rapid decrease in new technology product costs.

MAJOR ATTRIBUTES

The increased cost of the Smart Card buys important functional differences. These differences need to be considered in the economic justification of the Smart Card. Consider the differences between the FTC and the financial transaction Smart Card functions.

1. Capacity: The interchange Smart Card will routinely have 10 to 100 times the storage capacity of an FTC.
2. Connection: The Smart Card connection is a simple eight contact electrical interface. This compares with the magnetic surface of the FTC, which requires relative motion, a magnetics head, and supporting amplifier. Even with a manual card feed, the other devices are needed as is the added physical movement time delays.
3. Control: The PIN-based access control, the built-in activity and account limits, and the inclusion of available services and their

limits is a significant loss-avoidance and protection feature. The cost avoidance is achieved because the added protection allows use of the store and forward system attachment and operation mode. Introduction of the major information protocols enables the use of telephone-provided work stations rather than dedicated work stations. This is also an important cost savings.

4. Capture: The financial transaction Smart Card capacity to contain a transaction journal is valuable. It is an inexpensive way to avoid a labor-intensive or printer-based customer inquiry facility, providing there are enough details to satisfy the request. Of course, as the needed journal details increase so does the demand for the Smart Card capacity. Hence, this issue is a balancing act. Capacity versus required journal entries versus desired card life.
5. New system attachment modes: The remote and the store and forward modes might offer important cost savings. Elimination of dedicated lines and system units is one factor. Shared use of telephone and switched network facilities offers further extended coverage and reduced costs. This is particularly valuable as increased geographic coverage and interstate banking or interchange become a reality.
6. One hundred percent authorization: The logic and storage features of the financial transaction Smart Card offer a 100 percent authorized environment for all routine transactions. The exception and over-limit transaction will continue to require on-line authorization and control.

OPERATIONAL IMPROVEMENTS

There are a number of operational factors that offer cost reduction, cost avoidance, and productivity improvements.

Merchant Improvements.

Eliminate check writing delays at the checkout stand.
Eliminate the manual hot list look-up delays.
Increase credit availability by better control.
Achieve a faster checkout rate.

Financial Institution Improvements

Eliminate paper forms and data capture.
Reduce the expense of dedicated communications lines.
Eliminate printed hot lists and losses in printing and distribution time.

Achieve electronic data capture with the merchant providing the key entry function.
Increase resistance to merchant/customer collusion and fraud by better controls.
Increase the number of locations justifying electronic point-of-sale by reducing the associated expense.
Increase revolving credit income.
Reduce credit-line losses.
Provide uniform information protocols to encourage more cross-industry financial transaction Smart Card usage.

ECONOMIC AND APPLICATION JUSTIFICATIONS

The exciting part of the Smart Card is the new economic and application justifications resulting from Smart Card usage. Consider the three previous Smart Card examples (page 74).

The Increment Smart Card

This card offers a new cash and payment opportunity. It allows expensive, coin-based mechanisms to be removed. They are subject to theft and vandalism and a source of reliability and servicing problems. The payment card concept encourages the advance sale of multiple increments and allows and encourages self-service payment. It eliminates a coin collection, security, and auditing expense. If needed, each payment can have pertinent data (e.g., a ticket number) captured electronically for later reconciliation or analysis.

Social payments represent another use of the financial transaction Smart Card interface device. For example, it will accept food stamp collection at an electronic cash register. Thus the increment Smart Card can share the same acceptor unit as a cash payment device while achieving control goals. Eliminating paper, validating authentication, and guaranteeing no unauthorized reuse of the accepted increments are the added values.

The Interactive Smart Card Work Station

The inclusion of Smart Card functional capabilities extends this work station usability. It supports off-line station-to-station transactions and facilitates batch handling of financial transactions from an interactive Smart Card work station to the acceptor's bank without paper and using electronic data capture. The combination of Smart Card and work station logic introduces the benefits of the information protocols to improve Smart Card work station operation. The added identification, security, services and limits, and journaling facility enhances the work station. Use of dial-up facilities, reduced loss, improved inquiry into past transactions, and the ability to initiate financial payment and invoicing transactions from any telephone connection all contribute to the justification of the Smart Card work station.

The Interchange Smart Card

To appreciate the benefits resulting from the financial transaction Smart Card, one must first make a brief review of the current FTC. In the United States there are about 150 million issued bank interchange credit cards, which does not include the debit and ATM access cards. There is a significant amount of issue redundancy in the population using the 150 million cards. A typical household having interchange cards has two cards of one type. Also, they average two different cards per household. Given 150 million cards and using a household unit population of about 30 million household units results in about one-third of the households in the United States have cards. What about the other households? They don't appear to have bank credit card access. Why? They may not have asked for credit card access. However, more often they may not qualify because their income is too low. Or they may not have a sufficient credit record.

Credit card losses of all types average about 1.0 to 1.2 percent of the United States national gross credit card sales volume. Of these, about 85 percent are those losses that are due to a failure of card holders to pay their bill. In 1986, that was about $2.0 billion. Financial institutions are hesitant to offer credit facilities to those not meeting their criteria of credit worthiness. One important reason for this is the lack of effective control on the use of today's credit cards. About 50 percent of the locations at which credit cards are accepted today do not have credit card authorization work stations. Rather, they use the telephone to authorize transactions above a given floor limit. The floor limit restricts the number of transactions requiring a labor-intensive authorization process. Amounts under the floor limit are manually checked against a printed hot list.

The major new opportunity for the financial transaction Smart Card is twofold. One is to extend Smart Card issue to the other two-thirds of the United States households. That is a number of potential users, spenders, and credit users that is twice again as large as the credit card market of today. The average charge amounts for this group may be smaller, but the total credit need is more significant. The second opportunity is to reduce the cost of providing 100 percent control to more points-of-sale. The new system attachment modes are designed to reduce these costs. The information protocols are intended to give a very high level of control at all card acceptor points. The improved end-of-day batch handling of transactions also helps. The combination of functional advantages are designed to give 100 percent authorizations at three to four times the number of locations serviced.

The combination of three times the number of households and three to four times the number of authorization points is a significant additional marketing potential. There will be a great deal of competition for this market. The Visa electron card, described in the next chapter, is one competitor. The advent of very low cost authorization terminals will be necessary. These alternative solutions have some very significant problems, also discussed in the next chapter. The interchange Smart Card with its important information protocols has a significant opportunity to be the preferred solution. The successful solution to this extended market, customer base, and credit extensions represents a large added revenue and profit opportunity.

SUMMARY

All of the Smart Card configurations offer exciting new market improvements. Some are the classic results of migrating from paper- and labor-intensive operations to the modern benefits of electronic implementation. Others are increased numbers of users and applications. These result directly from the Smart Card attributes.

The other key economic factor is the decreasing costs of Smart Cards projected for the next decade. The continuous price performance improvement of integrated circuit chips is very significant, as it will encourage a serious look at the Smart Card by the financial services industry as an alternative to today's FTC.

13 Alternative Card Technologies

There is a very substantial worldwide investment in the magnetic-stripe plastic card. Why not continue to grow and extend its capabilities? Might there be enhancement features for the conventional card that would make it a more successful competitor to the Smart Card? Are there other technologies, techniques, or tools to match the Smart Card effort? If so, how might they be used? If not, then why not?

The alternative areas cover more than card technologies. Other system attachments, information usage options, and future technology options must also be considered.

EXTENDING THE MAGNETIC-STRIPE PLASTIC CARD

The magnetic stripe has several extendable dimensions. The various track densities could be increased, the number of recording tracks extended, or the recording codes changed to include other functions, such as an automatic error detection and correction code. Additional application functional allocations of recording tracks could be made: one or more tracks could be set aside for capturing a journal of card-based transactions.

Other new alternatives relate to protecting the stripe content. The recorded content of one or more stripes could be encrypted. This would prevent casual reading of the content for illicit purposes. I hesitate to suggest that encryption would prevent others from using the encrypted content. If not done carefully, encrypted blocks of known data might be exchanged for unknown data. The stripe technical characteristics might be changed to one requiring a greater effort to record or erase information. The stripe might even be made nonerasable, or the content may be tied to a permanent card property that is machine readable.

Assessment

A magnetic stripe by any other name has a basic set of traits. It is passive, its content is readable, erasable, alterable, or rewritable, and generally it requires moving the stripe under a reading or recording head. There are some exceptions, but the relative surface movement is the predominant technique. An increase to multiple tracks requires multiple heads, circuits, switching, and some form of location control. Density increases are limited. The current low density (75 bits per inch) and high density (210 bits per inch) recordings were selected to offer resistance to surface dirt and scratching characteristics. As densities are increased, the stripe readability becomes more susceptible to these external factors.

The theoretical prognosis is that the magnetic-stripe capacity may be increased to achieve the information capacity of the low-end interchange Smart Card. That added capacity will be costly, however, and the stripe will still be a passive device. It will lack logic capability and lack the local ability to journalize and provide security function. The encryption protection features have several costs. Each work station will need a key management and decryption function. This will add decryption execution time based on the amount of data to be protected, a costly feature if the national encryption standard has to be added and implemented in each work station. More details on these changes are given in Chapter 15 (see page 165).

Further density, logic, and feature growth will favor the Smart Card. In fact, the economic and functional break-even point might be reached with a five-year period. So that even while some added density changes will occur for magnetics, the solid-state Smart Card has too many unique advantages for which the magnetic-stripe and its on-line solutions will, at best, offer only a poor interim solution.

Several factors may accelerate the Smart Card solutions, such as a sharp growth in communications expense, or an alarming outbreak of magnetic-stripe frauds. These and several other changes would accelerate a close examination and a probable move to the Smart Cards.

CERTIFICATORS VERSUS AUTHORIZERS

These devices are intended to offer improved off-line functions. The VISA Super Smart Card (see page 92) builds an authorizer into the same package as the Smart Card. The Certificator (see page 135) is a stand-alone device that offers the similar control functions. Both alternatives are aimed at tapping Smart Card functionality at the low-volume usage location.

Assessment

Certificator advocates claim the device has at least a 5-year life, a low price, and may be shared by several card issuers. Built-in Authorizer advocates admit that the card costs are higher (e.g., $5.00). However, it adds functions to the Smart Card for which the card holder will be willing to pay. This includes a calculator, reminder pad, and frequent/important number memo pad. Authorizer opponents observe a shorter useful life (30 months average) for customer-carried cards, an added reliability problem due to many card-carried sensitive components, and an added cost for every card. If 100 cards are issued for each terminal location, that results in $500 in card costs (100 × $5), versus $50 for a Certificator.

THE DIGITAL OPTICAL "LASER CARD"

One form of nonmagnetic, high-density recording medium is called the "laser card" by its manufacturer, Drexler Technology. It is a special material that allows very high density bit recording in a record-only mode. An area the size of a signature panel can hold several hundred thousand characters, or about two million bits of information.

The recording area is reported to have a scratchproof covering. There have been no claims as to the susceptibility of the recording surface or reading process to dirt or surface contamination, so these could present a reliability exposure. There has also been no demonstrated or reported tested ability of the high-density recording to survive in the hands of a user population.

There are various ways in which the digital optical information could be protected from reading or alteration. The content could be encrypted, assuming the overhead execution time could be tolerated. The information when recorded is nonerasable, although it could be obliterated by overwriting. An area could also be set aside for a transaction journal.

Assessment

Laser cards currently cost ten times the cost of magnetic-stripe plastic cards ($2.00 versus .20). The laser-driven recorder/reader has a discussed cost of $500, a reader only at $300. And these are very unreliable estimates. However, these are still 20 to 30 times more expensive than a magnetic interface, which is, in turn, several times the cost of a Smart Card interface.

With projected card costs, the laser card and the financial transaction Smart Card will probably have the same costs within five years. However, it will take a decade for the Smart Card to approach the storage capacity of the laser card, if then. The laser card is a passive medium, it lacks any logical capacity, and its content is readable, overwritable, and rewritable. Hence, it has the same type of security exposure and response as the magnetic stripe. The lack of application and security logic are weaknesses compared to the Smart Card.

Overall, there are several applications in which the storage-only medium has significant advantage, for example, using the laser card for a large information store such as a medical profile or a program loading device or a hot card list. However, there is not enough of a need for such enormous storage to threaten the superior attractions of the Smart Card.

SYSTEM ATTACHMENT ALTERNATIVES

There is another way to achieve the projected results of the financial transaction Smart Card: put the whole world on-line. Each increment Smart Card would be a pointer to a record somewhere in a data base. Each interchange Smart Card would be a pointer to a user's data base in an issuer's data base. Each interactive Smart Card would be a simple display-keyboard work station pointing to appropriate information in a data base.

ASSESSMENT OF AN ON-LINE WORLD

In all three Smart Card examples, on-line access could substitute for the information protocols of the Smart Cards. This would create a very large and growing communications demand. Every use of a Smart Card would immediately create multiple messages to the issuer, to the acceptor, and to the interchange control system. This would be costly in communications expense and time. Other functions would be required, such as encryption of sensitive data in both directions. Programming and communications execution time would be needed. There would probably be a sufficient demand to require dedicated networks and equipment. Expanded network capacity would be necessary. Sufficient data base capacity would be needed to substitute for millions of Smart Card data bases.

Overall, the cost of on-line handling in a high-volume environment must be compared to the costs of the Smart Card solutions. Given a reasonable security plan and an explosion of communications costs to local point-of-sale units, the Smart Card might become economically attractive as card costs fall.

OTHER EMERGING SOLUTIONS

An area as attractive as transactional processing will always result in other solutions emerging. When the magnetic stripe was under serious consideration in the mid to late 1960s, at least half of the magnetic-stripe development effort was devoted to understanding and evaluating technological alternatives, which were being proposed faster than the selection committees could absorb them. My own approach was to never be surprised at any claim. And there were many claims. When a new proposal was made it was necessary to proceed through a point-by-point comparison with the main attributes of the base technology.

Personal Identification Technology

Several techniques are emerging in the area of personal identification: signature dynamics, retinal scanning, hand geometry, voice prints, fingerprint scanners. Each could be characterized as requiring large amounts of data, a broad range of comparison time, and as lacking any large-scale test with the using public.

A magnetic-stripe based solution will probably require a data base access for information required to make a personal identification. The information transmission would need to be encrypted for protection purposes that would prevent alteration or unauthorized interception and use. A Smart Card solution might have sufficient internal capacity for a signature verification and retinal scan. The others would need to fall back on the same solution as the magnetic-stripe card. However, as Smart Card costs fall and capacities increase, the Smart Card will have a growing advantage. In the long run, the Smart Card will be the more attractive user of the emerging personal identification verification techniques.

Public Key Encryption Algorithms

The public key is an encryption approach that uses two keys: one key for encryption, a second and different key for decryption. Hence, having an encryption key does not give the holder access to data encrypted by others using the same encryption key. Today, encryption is performed using the Data Encryption Algorithm (DEA), as discussed on page 131. It uses the same key for both encryption and decryption, so a key management process is required to give a separate key for each input user of a network.

The DEA protection plans have a fundamental need. Namely, to keep key values secret. For that reason, the Smart Card with a restricted access area is very attractive as a key management device. The Smart Card has an area that can retain a key and also prevent casual access to or inspection of that key. With the public key system, the key value for encryption need not be kept secret. However, for receiving purposes, each receiver would need to keep his key secret so as to prevent others from listening. This also represents a functional advantage for the Smart Card and its confidential access storage.

Secure Card Properties

As discussed previously (see page 36), one of the security vulnerabilities of the FTC is the need for a technique that ties the magnetic-stripe content to the physical plastic card. This technique resists frauds that replace stripe content independent of the plastic content. The stripe content might be from an acceptable card, while the embossing might be altered to an unissued embossed number. Thus, processing the imprinted paper would result in an unbillable transaction.

The same type of concern relates to how the magnetic stripe is tied to the chip for identification consistency. The financial transaction Smart Card will go through an extended transition period. It may be possible in this period for the chip content to be used for authorization and the stripe content to be used for data capture in an old work station. Again, the stripe content cannot be checked visually by the acceptor. A good authorization by the Smart Card content might result in a fraudulent capture record for an unissued number from the stripe. Hence, the transaction could not be cleared back to the Smart Card account.

The fraud issue is real. In fact, it may be equally challenging to the Smart Cards as to the predecessor cards. The proposed solutions of a nonerasable stripe segment with a unique number will apply equally to both card types. There is still some concern that the techniques currently being proposed are defeatable. However, I'll leave that argument for others to resolve. Suffice it to say, this area needs a workable solution. When found, it will be of equal value to current and future cards.

SUMMARY

There is too much invested in today's magnetic-stripe FTC to expect its early or abrupt demise. In fact, the current card still has several years of growth and evolution capacity. However, beyond these first few years, the capacity, security, and other features of the Smart Card will become very important. The continuing fall in Smart Card costs will further enhance its attractiveness. Thus far, there does not seem to be a serious alternative technology competitor.

14 Smart Card Standards

WHY FINANCIAL TRANSACTION SMART CARD STANDARDS?

There is a large market for financial transaction Smart Cards. They will be issued, accepted, and used in many countries, industries, and financial applications. Many suppliers and vendors will provide cards and their interfacing devices. Standards are a significant communications vehicle to assure that multiple suppliers and vendors will accommodate the Smart Card in an interchangeable manner. Hence, standards are an essential ingredient for worldwide and cross-industry interchangeability.

WHAT IS A STANDARD?

A standard is a voluntary agreement that is cooperatively produced for voluntary implementation and usage. The voluntary participants in creating a standard include the following participants:

Users
Providers
Industry associations
Consumers
Suppliers
Vendors

Regulators
Other standards groups
Liaison groups

Included in the standards process is an opportunity for others not directly involved to enter public comments.

The basic FTC standards are described on pages 22–23. There are a variety of standards. Some are based on information or data content (e.g., magnetic-stripe content), others are in the form of specifications (e.g., the magnetic-stripe recording location on an FTC), and some relate to materials and their characteristics (e.g., the magnetic-stripe material characteristics).

WHAT ARE THE STANDARDS-MAKING BODIES?

There are many! There are perhaps hundreds of groups in the United States involved in drafting standards. For the financial services industry, the standards-drafting bodies include the following groups:

Leading users
Large financial institutions
Industry associations (e.g., National Automated Clearing House Association)
American National Standards Committee For Financial Services (ANSC X9) (This is also the United States agent for International Organization for Standards participation.)
International Organization for Standardization Technical Committee for Banking (ISO TC 68)
National Bureau of Standards
Consultative Committee for International Telephone and Telephony

HOW ARE STANDARDS CREATED?

There is no one way. Some efforts start with a leading user, then migrate through industry working groups to an American National Standards Committee working group. Others start with vendors and migrate through a few users to a technical committee working group. Others originate outside the United States and migrate to an International Organization for Standards working group, which are then carried to national activities.

Generally in the United States, the American National Standards Committee for Financial Services Standards (ANSC X9), focuses on financial industry standards needs. Card technology is handled by ANSC X3B10, the card technology subcommittee for the Information Processing Technical Committee (ANSC X3). This group is the focal point for the cross-industry "identification card" standards. The current financial transaction card standards is their creation in cooperation with the ISO TC 97/17 subcommittee. The Smart Card activities are under the same Technical Committee. The responsible Smart Card working groups are ISO TC 97/17/4 in conjunction with the United States working group (ANSC X3B10.1). Currently, the leadership role is in the ISO working group. The French standards organization, AFNOR, provides the leadership for the ISO 97/17/4 integrated circuit card working group.

WHAT IS THE CURRENT STANDARDS STATUS OF THE FINANCIAL TRANSACTION CARD?

The status of the standard FTC was reviewed in Chapter 3. The responsible international standards subcommittee (ISO TC 97/17) has just completed a major updating and modularization of the identification card (type 1) standards. (The main components were listed on page 23). These are worldwide-accepted standards for the financial transaction cards. There are several groups that act to protect portions of the standards that are of particular value to their industry. The airlines industry generally watches track 1. The banking industry tends to watch track 2, as do the vendors with equipment in the market that uses this stripe standard. Track 2 is, by far, the predominant usage area. The smaller financial institutions keep a watchful eye on track 3.

Track 3 is the only rewritable portion of the magnetic-stripe standard. It is used by small financial institutions to support off-line cash dispensers. It does this by writing the updated customer controlling data base back onto the third track after each transaction. The larger financial institutions have felt this to be a serious fraud exposure. Anyone with a rewriting capability could, in theory, modify the decision data base on track 3 to their advantage without bank control.

THE OTHER STRIPE STANDARDS SETTERS

In creating the magnetic-stripe standard, the standards drafters left a significant unused recording area for discretionary information. This has become a useful area for the major interchange organizations such as VISA and Mastercard. These groups have created several organizationally oriented standards that have been helpful in their operational processes. For example, VISA has added a 5-digit PIN validation field on track 2 in this area. This field is designed to allow VISA central to make a PIN validation when the card issuer system is not available for a decision. VISA central gets the issuer's permission to provide that service. It is a DEA-based process with VISA central maintaining a few secret encryption keys for each participating institution.

The track 2 content is getting close to being full at 40 digits including a few control characters. There is some preliminary finance industry discussion about migrating from track 2 to a banking industry format for track 1. That would give the finance industry up to an added 60 digits. However, the first 19 characters would be common for both track 1 and track 2. If that move is needed and made, there would be a notice served to the industry and to vendors to prepare by a specified date to read both track 1 and track 2 at financial services work stations. This move would not reduce any of the conventional security dangers discussed previously and would not offer any of the information protocols of the financial transaction Smart Card. In short, the move of the finance industry to track 1 would be an interim move. More on these changes is given in Chapter 15.

STANDARDS STATUS FOR THE INTEGRATED CIRCUIT (SMART) CARD

The initial work on Smart Card standards occurred in the French Standards Organization (AFNOR). They focused on the financial transaction Smart Card, and prepared a proposal to the International Organization for Standards Technical Committee on identification cards (ISO TC 97/17). The proposal suggested the formation of a working group to draft standards for the integrated circuit card with contacts. Several major card-issuing countries supported the suggestion (France, Germany, United Kingdom, Japan, and the United States). The ISO working group (ISO 97/17/4) was then formed with the objective to draft an international standard for the financial transaction Smart Card.

The basic FTC standards documents were rewritten recently in a modularized form:

Part 1: Physical characteristics

Part 2: Dimensions and location of the contacts
Part 3: Electronic signals and exchange protocols
Part 4: Minimum commands

These modules are now being drafted and circulated for the initial vote as draft standard. The drafts will complement the modularized standards for the financial transaction card. The four-part drafts cover the description of the financial transaction Smart Card.

Physical Characteristics—Integrated Circuit Card with Contacts

Those elements that are unchanged from previous standards (see Identification Cards—Type ID-1 Physical Characteristics, ISO 7810) are:

Materials
Dimensions
Deformation
Inflammability
Toxicity
Resistance to chemicals
Environmental temperatures
Humidity
Durability

Those elements to which additions have been made for the Integrated Circuit Card with Contacts (type ID-1 cards) are:

X-ray: A cumulative dose per year is specified that must not damage the chip.
Planar definition of the card contact set must not exceed a specific maximum.
Mechanical resistance of the card in normal use will stay intact.
Sealed surfaces will remain unbroken.
Contacts will not interfere with the magnetic stripe or its interface mechanisms.
Mechanical pressure must be sustained by the contacts.
Electrical resistance maximums is specified for the contacts.
Electromagnetic interference to be tolerated between the magnetic strip and the chip is specified.
Electromagnetic field sufficient to erase the stripe will not cause the chip to malfunction.
Static electricity discharge will not cause the chip to malfunction.

Heat dissipation of the chip and the surface temperature maximum is specified.

There is an addendum that is not part of the standard, but does provide information about bending, torsion, and static electricity.

Dimensions and Locations of Contacts

The second part of the draft standard covers the dimensions of the contacts, the number and arrangement of the contacts, and the location of the contacts relative to the upper-left corner of the card front (Figure 14.1).

Electronic Signals and Exchange Protocols

The third part of the draft standard relates to the signals and protocols between the integrated circuit chip and the interfacing work station. There

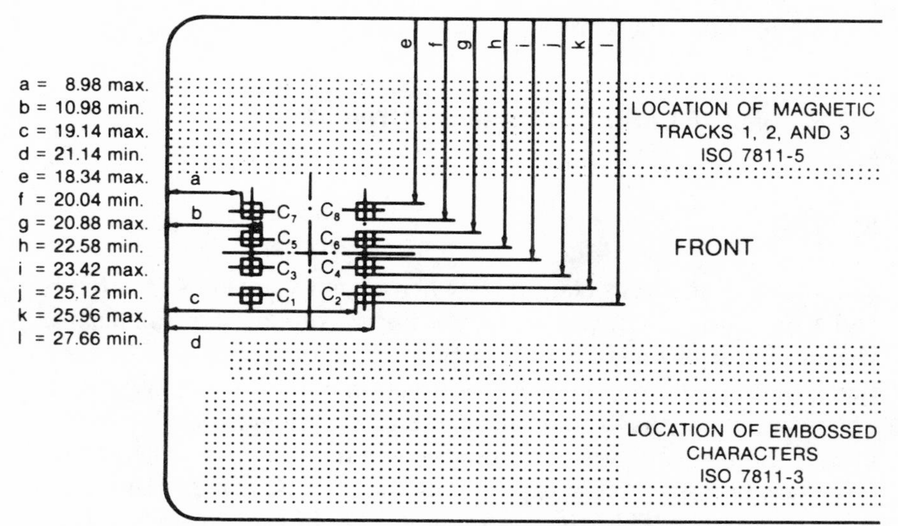

Figure 14.1. The draft standard for FTC contacts. Looking at the card front, the contacts are arranged in two columns in the left half of the card. They can be located on the front or the back side of the card at the issuer's choice. The minimum and maximum dimensions are given, respectively, in reference to the outer and inner rectangles defined in ISO 7813, section 5.1. This international standard defines 8 contacts referred to as C_1 to C_8. The contacts C_7 and C_8 are provided as unassigned contacts for additional use in certain applications. (ISO TC 97/17/4, Document N-82, Dated 84/03/28.)

is a possibility that this module will be split in two, one for signals and the other for information content. The present contents are:

- Assignments of the contacts
  - Function
  - Voltages and currents
- Starting operations with the card
  - Power on
  - Reset
  - Data exchange
  - Setting voltages
- Transmission mechanism
  - Modulation rates
  - Transmission cycle
  - Parity checking
- Communications organization
  - Answer to reset from the card
  - Information formats
  - Procedure information to the terminal
  - Information transmission

OTHER SMART CARD STANDARDS ACTIVITIES

INTAMIC

The International Association for Microcircuit Cards (INTAMIC) has had a standards activity. This organization is composed of financial service organizations in several countries with a joint objective of helping the evolution of the financial transaction Smart Card. Within the organization they have established several working groups. For a while, one of these groups was examining the standards area. They plan to introduce, from time-to-time, proposals into the ISO 97/17/4 Smart Card working group. INTAMIC does maintain a liaison function with the working group. That means that someone attends the working group meetings for coordination purposes.

Currently, the INTAMIC study groups are focusing on these issues:

Vocabulary
Interface on the physical level

Interface on the logical level
Proposals for physical and logical security

The following areas will be considered for common definition activities:

Application processes
Housekeeping functions
Card control functions
Card manufacturing control

Hopefully, these efforts will provide added input to the ISO working groups. The ISO groups are the most likely source of further standards results. It is an open forum, has an excellent start in the three drafts underway, and has the broadest scope of participation.

Standards Status for the Banking Industry

These latter INTAMIC efforts relating to applications and security have been proposed as Banking Standards. INTAMIC has made a proposal to this end to ISO Banking Standards (ISO TC 68). These have been approved as items for two working groups (ISO 68/6/5 and 7).

The application effort will initially start with a focus on interchange data (ISO 68/6/5).

It is intended to spell out the minimum set of interchange data to be provided during a Smart Card transaction. I would expect this to include, as a minimum, the same set of fields provided by the magnetic stripe. This relates to any track (1, 2, or 3), but it will most likely relate to track 1 or 2. The fields included are:

Issuer industry
Issuer identification
Account number
Account number check digit
Expiration date

There are several other optional fields on the magnetic stripes. I would expect that these added fields would also be available in the data from the Smart Card.

The security work item has been outlined. It will focus on the security groundrules for the card life cycle and card based transactions.

Other Industries

There are a number of industries that will be using Smart Cards: the telephone, medical, government, manufacturing, and entertainment industries. The telephone industry has several standards-setting bodies for technical standards. It is unlikely that the technical groups will step up to these issues. Rather, it will fall to end-product groups such as the videotex users or servicers to cooperatively prepare the appropriate standards. They will, most likely, follow the pattern established by the ISO working group for the financial transaction Smart Card. The government and manufacturing industries will also follow the ISO pattern. Many of their cards will be used for payment and invoicing activities. Hence, a compatible card will be worthwhile. Other industries with unique needs will still start with available Smart Cards whenever possible. This will let them get the economy of scale and solution know-how developed in other application areas.

The medical area poses a challenge. Assume that a universal medical profile Smart Card definition is initiated. It will probably occur in several countries simultaneously. I don't believe there is an interest or motivation or mechanism for those groups to collaborate, so if there is to be a cooperative effort I believe it will occur in a government or insurance area. Then the result can be "standardized" by the ISO TC 97/17/4 integrated circuit card working group.

OTHER STANDARDS ISSUES

Beyond the immediate standards world of the Smart Card and the financial transaction card there is one other: the standards world of communications, information processing, banking, and programming languages. Several of these areas are worth a brief review in order to understand their relationship to the Smart Card. Some represent areas to which the Smart Card will interface. Others will benefit from Smart Card considerations.

Communications Protocols, Architectures, and Services

There is a wide variety of standards, draft standards, and standards debates in this area. They range from analog to digital, from asynchronous to synchronous, from conventional to new alternatives, and from conventional instruments to radically new delivery vehicles. Most of these will be chosen by the network and system designer. The role of the Smart Card will be in the job and device card roles with the work station. These cards will need to anticipate the interface and standards details. From the Smart Card user's point of view, these issues will be transparent. That is, the Smart Card user will not even recognize that an appropriate device card has been inserted into the work station. In many applications, the user may not even recognize that a communications session was required and completed. The new attachment modes of remote and store and forward will need to be carefully structured. The structure will need to achieve the correct communications mode and protocol, as required.

Higher-Level User Programming Languages

Programming languages are not currently a requirement in Smart Card use from the user's point of view. The Smart Card user does not use or interact with these languages. The Smart Card user deals with menu selection. The internal Smart Card programming has been done by the developer. More likely than not, the developer has used very sophisticated microprocessor languages and tools, designed for maximum performance and maximum use of minimum Smart Card chip capacity. These languages are generally not the basis of user standards; rather, they are provided by the chip supplier to the developers.

File Structure and Access Languages

The Smart Card user also does not deal directly with data base access and structure issues, but with option menus. The internal logic of the job and device cards interfaces with the internal information processing structure. The Smart Card in use will be sharing data base accesses. A host-based, issuer-provided, single-customer data structure mapping is an essential ingredient in the evolution to electronic transaction systems. The data structure need not be all in one location, but it does need to be mapped and accessible. Hence, this is an important area for issuer-used standards, common structure, shared mapping and easy access. Once again, this is a total transparent area for the Smart Card user.

Open Systems Interconnect

This is a developing reference model for a common intersystems connection. It is defined in seven "layers." The top most layer is the applications interconnection portion. This is essentially a pipeline through which the application specifics are dealt with by the interfaces. The open systems interconnect definitions do not provide any standardization within the applications layer. *This is an important opportunity for the Smart Card to contribute to the open systems interconnect definitions.* The information protocols of the Smart Card are an important and definitive need of a common applications structure. The identification, security, services, content, and journalizing structure of the Smart Card could well be a common transaction application interface. Even if the open systems interconnect standards activity became aware of the Smart Card Information protocols, it would take several years for this proposal to get serious standards attention. It is offered here as an observation and opportunity. It should not be an issue to delay early Smart Card ventures and developments.

Magnetic-Stripe Content

Magnetic-stripe content is an important standards area for the Smart Card. The financial transaction Smart Card must coexist with the conventional financial transaction card. The two cards will be serviced by the same systems, data bases, and networks. The three magnetic stripes (tracks 1, 2, and 3) form a minimum common data requirement:

Primary account number	19 digits
Expiration date	4 digits
Restriction or card type	3 digits
Offset or PIN parameter	5 digits
Total	31 digits

The last two fields are shared by tracks 1 and 2. These data elements leave only eight digits of discretionary data for track 2.

The financial transaction Smart Card will need to have these data items in common with the magnetic stripe. Beyond that starting point, the financial transaction Smart Card has greater capacity and function. Hence, a significant standards effort will be needed to identify, size, storage, designate, and secure the Smart Card content. There are some that would consider this a discretionary area, who would leave it to the individual application development group to set up and allocate the remaining application data needs. In any event, the issue will be examined by the ISO 68 standards group. This effort has just been initiated in the ISO 68 Banking Standards Committee: the early financial transaction Smart Card innovators will still probably have a free reign in this important question.

Personal Identification Number (PIN) Security

There is a significant standard for PIN security and management (ANSC X9.8, 1982). The Smart Card implementers will need to understand and build these standards and guidelines into the security process and the information protocols. The PIN security standards prescribe the source of the PIN and all of the steps relating to its full cycle of events and life. It also addresses the manner in which the PIN is routed through the implementing system. It is essential that the Smart Card implementers observe this standard. This will assure the coordination of a common electronic systems evolution with all card-issuing nations. It offers the user a common PIN base with his various electronic services needing a PIN-based access.

Message Security

This application area has recently had a United States standard issued (ANSC X9.9, 1982). This is also being advanced as an international standard through ISO TC 68 (Banking). The standard addresses an encrypted check value for wire transfer and other important messages. This standard will be a common tool for financial transaction Smart Card holders in the area of network and international banking services. However, this standard will be implemented as an internal action. It will be transparent in response to a menu selection by the Smart Card user. There will be added standards developments in this area. Key management and message-protecting standards are certain to evolve. Hence, these too will need to be transparent to the Smart Card user.

Card-Initiated Messages

A major international effort, on a cross-industry basis, is currently concluding a standard in this area (ISO TC 68 and TC 97/17). This too is an implementation transparent to the Smart Card user. The application developers will format the Smart Card transaction message implementation to use these important agreements.

SUMMARY

Standards are generally a poorly understood and followed area, but most efforts are producing useful and meaningful tools. The wide scope of the subjects discussed in this chapter demonstrates the large number of important areas that need to be understood and accommodated in future Smart Card development plans, and shows that there are some gaps in the current standards activities. The scope of the Smart Card industry extends beyond the ability of any one group to dictate or achieve the needed standards. The sooner the participants (1) recognize the opportunities for cooperation; (2) join a meaningful and recognized national or internationl standards activity; and (3) contribute substantially, the sooner a complete spectrum of practical Smart Card standards will be available.

15 Magnetic Stripe Evolution or Smart Card Migration?

THE MARKETPLACE SHAPES TECHNOLOGY

As often happens, the choice between technological alternatives rests with the marketplace. Although there are important technology questions needing answers, FTC market forces are changing and need equally to be understood.

The original market concept for the magnetic-striped FTC was as an inexpensive, machine-readable, on-line "pointer." It pointed to the customer's data base and card activity accumulator in an on-line system. The magnetic stripe in this case was only a substitute for a customer-entered pointer, or primary account number (PAN) of nineteen digits. Why? Human-factor tests indicated that most users could not quickly or accurately manually enter the nineteen-digit number.

In the late 1960s, when the magnetic-stripe FTC was still new in the marketplace, knowledge of the card's potential and it s liabilities was limited. But the market served by magnetic-stripe FTCs has changed significantly in the intervening years. The market will change even further by the late 1980s, and that ought to be today's planning target. By then a full two decades of experience will have been accumulated. What is known now about the marketplace forces? Many changes have affected the environment of FTC use, adding to our experience and knowledge.

Skilled attackers and known loss categories
Use of complex networks, sharing, interchange, third parties, and other communications arrangements.
Government regulations for card liabilities, receipting, and usage constraints.
Card use in a wide variety of work stations.
Pin-based access in addition to that for ATM use, e.g., point-of-sale.
Access and security solutions in home, telephone, personal computer, and other remote banking environments.
Cross industry (e.g., airlines and retail) and multinational acceptance of FTC-based, unattended transactions.
Escalation of communication costs and intense development of numerous alternative paths and control points.

These changes are creating new FTC function and use demands. What are these demands? How will card content change? How do the Smart Card attributes compare with emerging and future needs? Should there be an abrupt change to the Smart Card, a transition phase, a plan to keep both alternatives, or only the magnetic-stripe FTC?

FTC CHANGES

Three types of changes in the way the FTC functions are now projected to match current and projected market needs.

Security: The original magnetic stripe content was unchecked. In fact, its content is completely changeable, alterable, copyable, and susceptible to several forms of easy fraud and losses. This is no longer acceptable. Protection against physical card alteration and counterfeiting, stripe content alteration, and stripe content copying are required. In addition, an improved form of personal identification verification (PIV), such as signature dynamics, is desired to eliminate the need for PINs and to decrease customer resistance.

Data Base: The original magnetic stripe content pointed to the on-line data base from which the transaction was authorized. The current demand for fast transaction handling is still on-line. However, the demand is for stripe-contained data that facilitates the transaction through work-station logic. The local decisions are then conveyed to the on-line center. This two-part process of local decisions and message confirmation speeds the transaction and avoids unacceptable customer-sensed delays. Included in this local decision set is PIN/PIV validation, controls based on permissive services and limits, national versus international usage, and acceptable account relationships or card account types such as debit, credit, and/or access.

Operational Concept: The original magnetic stripe content assumed a direct on-line attachment of the accepting work station to the controlling computer center with appropriate application and data base access. The current and future communications environment is much more complex. Network and center identification, accepting or backup nodes, transaction routings, public-switched network, third-party network, and interchange or shared facilities routings are all alternatives that need to be used. Appropriate security, key management, and possible work station mode options are all new variables that need to be anticipated and accommodated. The new and extended magnetic stripe content will need to reflect and respond to these additions.

PROJECTED MAGNETIC STRIPE CHANGES

Achieving the projected functional changes will require a series of magnetic stripe content changes over the next several years. The first of these changes, described below, has now been committed to by the major card servicers. The next three are under active design and market review. The last two are starting to appear in proposed national standards changes.

The VISA MCAS (membership-controlled authorization service) combines several of the following changes in a migration to track 1. This combines PIN control (partial), crypto check digits, and an off-line authorization limit.

My projection of the magnetic stripe FTC changes through 1990 are:

FUNCTION	CHANGES
Inhibit physical card changes	More sophisticated graphic arts content
Establish a card/stripe relation	Add a new track zero "Watermark;"® (EMI/Malco)
Increase basic track capacity	Extend FTC use to high-density track one
Protect stripe(s) content	Add stripe content encryption and track zero (WM) tie-in

FUNCTION	CHANGES
Add a PIV validation capability	Add a new track (TN1) for PIV validation (e.g., 66 bytes for a signature-dynamics data base)
Add multiple relations identification	Add a new track (TN2) for transaction, investment and savings accounts identification (e.g., 19-digit PAN for each relationship).
Add communications identification, routing and protection data	Add additional data to the above tracks

These projected changes go well beyond a simple addition of data to an existing magnetic stripe track. The multiplicity of changes includes:

New track "geography" on the back of the FTC
New track heads
Added space in the reader for the enlarged head, amplifiers, and power
Extended buffer space and logic to manage multiple tracks and content
Extended application logic to support work-station-based decisions
Increased message lengths to accommodate added content
Addition of encryption function, key management, execution time, and physical protection of the devices and sensitive data
Increased central data-base and application logic to recognize the various stages of work station evolution

RESULTS OF MAGNETIC STRIPE CHANGES

Card-based systems are not inexpensive. Large card volumes, frequent card reissue (e.g., every one to two years), existing and projected work station populations, and supporting administrative expensive are cost elements in every system. To these expenses must now be added the cost of the projected changes. The projected changes will be expensive:

Basic card costs will increase from 2 to 4 times for addition of the above changes. For example, the first two changes will each raise the cost of a raw or prepersonalized card by 50 percent or for a total of about 125 percent above the basic card cost of today. Total raw card costs will increase from 0.10 to $0.40. To this, add the personalization and delivery expense, although these costs will increase for all card types.

The cost of a magnetic stripe reader will increase by 1–200%, to $1500 from today's $500 reader. This increase includes up to four more tracks, packaging, power, card alignment for reading, and multiple density readers.

The logic changes in the reader and work station will be extensive. Watermark validation, head/track/density recognition, content projection, key and encryption management, service controls and limits, and message formatting and network routing controls will be needed.

The system changes include message additions, data base extension, logic upgrading, application and system development, and change management.

The extensive additions to card tracks, controlling logic, and data content will increase the complexity of system and application development. The additions will also increase the usage complexity. Which generation card am I accepting? How does it go into the work station? Is this the correct graphics design? Within the industry, standards issues will abound. Who controls which track? In which standards activities does the bank participate? Which vendors will support my decisions?

Not only will the projected FTC changes be costly, but they have another more difficult, and disturbing characteristic. The likely changes to the magnetic-stripe Financial Transaction Card will fall short of the market needs. Limited capacity, interface expense, incomplete security, and continued use of a mechanical solution all contribute to an inadequate end result.

Consider some of the following observations of a projected multi-striped FTC solution:

Five stripes offer up to 400 digits of capacity, but the projected need is for several hundred digits of storage:

Multiple relations: 3 relations × 100 digits each	300 digits
PIN/PIV/acceptable bank and personal passwords	200
Network and routing designators	100
Total	600 digits

The five magnetic stripes have little added room for the unexpected. Further requirements growth would need additional stripe tracks and geography on the FTC.

Control: The FTC stripe security facilities are limited. Even if the content was encrypted, accepting organizations would need access for local decisions such as PIN/PIV validation. Thus, the accessing organization could see the entire stripe, its sensitive content, and important control values. This would make customer data subject to privacy invasions or use by potential competitors to their advantage. To get access, the magnetic-stripe FTC system would need key management facility, workstation encryption and decryption functions, and the necessary time to perform the conversions. The work station would need to provide the appropriate protection of this sensitive process and its vital values. In any event, even with protected content, the stripe would be easily changed, copied, or altered, deliberately or by accident.

Market Use Control by the Acceptor: A related concern with magnetic stripe security is that all functions are achieved through the acceptor's work station. This puts the card issuer at the mercy of a device and logic over which he has little or no control. Such transactions would be uncontrolled until after the transaction has been reported. Perhaps 50 to 60 percent of today's transactions are authorized locally. Thus a rash of under-floor-limit transactions would not receive any issuer review at the transaction time. Future issuers will insist on a sampling of these events. The magnetic-stripe content could not force such an on-line sampling till there was industry-wide agreement to the revision of the work station authorization or interchange rules.

Mechanical Technology; The magnetic stripe FTC is a 40-year-old electromechanical technology. The recording technique was developed during World War II. The change to multiple tracks and high densities will cause technology complications. For example, the use of low-cost, single-track slot readers is probably not feasible. How does the acceptor designate which stripe track to read during a transition period? If an updating of stripe-based control fields is needed, a rewrite operation would be required. Misalignment of the card by the unit or the user might erase an adjacent track, which would not be detected until much later. Stripe reading accuracy would still be subject to surface dirt, marks, and stripe alignment, aggravated by a change to higher density recordings. Unfortunately, electromechanical devices have not had the price-performance and reliability improvements realized with solid state and integrated circuit chip devices.

CARD NEEDS ARE EXPANDING

Future changes of the magnetic-stripe FTC are significant. However, they are only a part of the changes to come.

1. Additional data content to the FTC for transaction-point decisions is needed in an on-line environment. (This capacity was outlined previously.) This data would provide a number of important results:

Minimize on-line messages and responses required for each transaction, minimize delays during the transaction.
Maximize communications path alternatives and the types of work stations that can be used.
Increase customer acceptance and use through better card function and control, such as use of a PIV instead of having to memorize a PIN.
Reduce losses and misuse through better security options.

2. Uniform exchange of data between card and work stations is vital to achieving wide card use and acceptance. A uniform and consistent interface protocol for all remote transactions is needed for the following elements:

Identification: Individual, issuer, and all relationships
Security functions and access controls
Service options and limits
Account balances, usage, limits, and resettable time periods
Message routings, network identifications, and protection requirements
Personal identification options and validation data (PIN and PIV)
Customer and financial institution personalization, acceptance codes, and usage options
Usage journaling

3. The logic for card transaction acceptance and usage control should be shifted from the acceptor's work station to the issuing financial institution's FTC, by including usage control logic and data as a function in the FTC. The logic is then executed within the FTC. This change is designed to prevent acceptor's confusion resulting from a growing number of different card types, data content, and required controls. Given the large number and variety of accepting work stations in a number of interchanging systems, there are a number of card-based facilities needed:

Card-based logic control and controlling data for application, services, and security implementation
Card acceptance and validation of PIN and PIV parameters
Card-dictated, external-to-work-station, on-line authorization
Card-specified message routing, content, and protection
Card-captured transaction usage journal
Card-controlled account and services access alternatives and controls

THE INTEGRATED CIRCUIT (SMART) CARD

Let's review the Smart Card attributes before assessing the answer to evolution versus migration.

1. Capacity meets the expanded FTC needs: The Smart Cards being offered in the marketplace have capacities that exceed the projected needs. Early card entries now have 64 thousand bits. In addition, they offer complete internal control of the storage access in several modes of access. Included are free access, limited access, and externally unaccessable storage areas. The content of the card may be personalized to each customer's and financial institution's needs. Several areas are also selectively resettable by the customer and the financial institution. A major provision is for a permanent journal of card activities. This is very valuable as customers participate in an increased number of self-service and unattended operations.

2. Control Logic Internal to the FTC: Added to the storage capacity of the Smart Card are internal logic facilities. This facility returns important control capabilities to the card-issuing financial institution. For example, the card logic can decide on the need for external, on-line authorization based on card activities. This could be an important advantage over clerk-exercised, floor limits. The internal logic also reduces the demand on work station function and complexity. The logic is set by the card issuer, executed within the card, and has access to the important control and activity data within the card. By including key-driven algorithms and keys within the card, the key management issue and complexities can be avoided in the work station and its supporting system. The internal logic returns control of card activities to the issuer. An important new attribute is the ability to remove the primary account number from the external card data. When the proper PIN/PIV value is presented to the card, it will respond with this data to an on-line transaction. Thus, on-line control is assured for a prescribed customer set, such as marginally qualified credit customers.

3. Connection Ease and Simplicity: The Smart Card is a solid-state,

electronic device. It is read and interacted with on an electronic basis through a standard eight-contact and electronic signal protocol. The protocol is capacity independent and transparent. Thus cards with a wide range of capacities would present a common interface by adjusting to its internal capacities. For example, it would know when its activity journal was filled or functional limits would prevent transaction participation. An important attribute of a solid-state device is that it continues to have a very impressive ratio of price-to-performance improvement.

4. Transition is built into current standards: The international standards (ISO 97/17/4) for the FTC Smart Card provide for full compatibility with the current physical, embossed, and magnetic-stripe FTC. Thus, appropriately manufactured FTC Smart Cards will work in conventional embossed and magnetic-stripe work stations. This allows the investment in these earlier work stations to be projected for the normal life of the machine. Of course, the earlier work stations would need to be updated if the additional attributes of the Smart Card are to be used.

THE SMART CARD IS A BETTER FTC ANSWER

The Smart Card limits or stops the dependency on magnetic-stripe patching: The magnetic stripe is an old technology. As an electromechanical technology, its expansion and change will be costly. It will continue to be expensive as future desire for change arises in the marketplace. It lacks the ability for any substantial improvement capacity without an upfront investment. Hence, further changes will be delayed until the inertia is overcome to put an industry patching-committee in place. Clearly the solid-state Smart Card puts the marketplace evolution back in the hands of market-smart financial product producers. There will be areas in which the magnetic-stripe FTC will continue to be a more economic, if limited, solution. The Smart Card allows that joint market solution. Its compatibility makes that possible with full transparency to an accepting clerk or merchant.

The Smart Card attributes best satisfy future FTC needs: The Smart Card functional and capacity characteristics are an excellent match to the projected requirements. Now is the time for the financial industry to tap the outstanding and continuing price and performance improvement of the Integrated Circuit Chip. Chip economics, capacity, function, and performance are still on a sharp growth improvement curve. The chip is well established as a fundamental technology building block. Each of us routinely carries and uses chips on our person. Chip-based watches, calculators, appliances, radios, TVs, telephones, and automobile elements put aside any concern of useability, survivability, and effective price performance.

Adoption of the Smart Card as an FTC puts the card-issuing institution back in control of its card usage: The content, controlling logic, and applicable data base are established by them. The logic is executed within the card when desired by the issuer. Criteria for external actions may be controlled from the card. Vital data protection, externally unaccessable data, and the protection techniques may be controlled by the issuer-set logic and controls. In addition, each card has a permanent record of its use that can be accessed when appropriate. Of course, both the customer and the merchant also receive paper receipts for backup and verification purposes.

The Integrated Circuit (Smart) Card is an economically justifiable financial transaction card, and its cost is falling rapidly. Just a few years ago they were described as being in the $40 to $50 range, depending on country of origin and quantities—which were quite small at the time. Today they are only about 4 to 5 times the cost of a *delivered* magnetic-stripe FTC, and a possible 2 to 3 times the cost by the end of the 1980s seems realistic. That is a very justifiable target now, for these projected benefits:

OBJECTIVES	ICC FUNCTION	ECONOMIC VALUE
Reduce losses	Built-in logic, data, and controls	Better control, improved on-line-sampling of floor limit criteria
Increase active cards	Built-in PIN/PIV validation	Customer motivated to use cards having quicker and easier functions
Better operating costs	Built-in logic, data, and controls	Reduce messages required for on-line operation
	Built-in PIN/PIV validation	Shift function to card, reduce work station cost and changes
	Built-in key management and algorithms	Improve security at lower cost with portable key-management device
Increase revenue	Built-in logic, data, and controls	Increase customers eligible for credit by better controls
Develop new services	Built-in data and controls	Introduce a cash card to reduce cash-handling expense, increase float and cap ATM needs

These economic factors will produce a range of savings. The savings are a function of local assumptions regarding cost factor sizings, cost of funds, and values associated with market results. In actual projections, these factors have produced savings in the range of 5 to 10 times the cost of a delivered Smart Card.

But there are some other important economic considerations. For example, in most financial institutions, there are 3 or 4 magnetic-stripe FTCs issued for each active card. That suggests that Smart Cards should be issued to customers that have an acceptable FTC usage record. This avoids or minimizes the expense of unused Smart Cards. It also creates an added revenue option. Namely, that a customer who wants a Smart Card prior to establishing their usage record could be invited to request a card at a reasonable card fee. Thus, one Smart Card could avoid 2 or 3 conventional card issues and also earn a customer-paid fee. That arrangement makes the functional and market advantages of the Smart Card even more attractive to the issuing institution.

EVOLUTION OR MIGRATION? BOTH!

All evidence suggests that the magnetic-stripe FTC will have a place in the future. A financial institution with a static market, a significant investment in magnetic-stripe work stations, a very low card acceptance rate and/or rapid customer turnover, and little prospect of additional types of electronic services will probably stay with the magnetic-stripe FTC. This would not penalize their customers for several years.

At the other end of the spectrum is the institution with a large stable of aggressive magnetic-stripe FTC users, a fast-growing range of electronic services, an increasing set of interchange and sharing arrangements, and a growing concern about magnetic-stripe-based losses and frauds. That institution will take an early look at Smart Card technology. Internal testing, selective field tests, preliminary product development and options identification, and active industry experience pursuit are all important moves.

In between these two extremes is the majority of institutions. An early designation of a responsible product identification and market assessment executive is appropriate. This product area is characterized by three important customer types: mobile, young and professional, and innovation acceptors and participants. If the shoe fits, then prepare for early assessment and decisions.

In the final analysis, an active effort to accommodate both types of financial transaction cards appears to be the appropriate action path. May your circumstances afford you the luxury of careful and considerate actions based on a body of thoughtfully prepared and gathered evidence. I never seem to achieve that luxury!

16 Societal Impacts

LONG-RANGE CYCLIC CHANGES

Believers in long-term, repetitive cycles suggest that the 1980s are the transition period to a new fifty-year cycle in United States society. This transitional point in the cycle has a number of predictable social characteristics:

Changing role of government
Energy base transition
End of transportation growth
High technology turnover
High capital obsolescence
Societal occupation shifts
Nationwide financial, budgetary, and deficit problems.

These factors, which coincide with a fifty-year cycle transition, could be coincidental. However, the issues are real! Furthermore, the displacements occurring in our society are quite broad. In fact, there is some concern that the induced societal transitions may be growing. Consider the following growing concerns. Are there:

Too many factory workers for robots?
Too many steelworkers for the third world steel mills?
Too many secretaries for word processing?

Too many tellers for ATMs?
Too many letter carriers for electronic mail?
Too many programmers for the new end-user languages?
Too many labor- and paper-intensive workers for the emerging electronic solution alternatives?

MIGRATION TO A NEW ELECTRONIC ERA

The impression one gets is that there is a need to shift emphasis from a smokestack economy to an electronic information society, a need to migrate an undertrained, undereducated, and obsoleting smokestack work force to a new era. It will be a transition marked by a contradiction. It is an era in which vast numbers of people who are untrained and possibly untrainable in information processing must become capable of using information processing as a high-productivity tool.

The potential for acceptance of new technology at high speed by untrained personnel has been demonstrated before. The important ingredient was packaging the new technology in a form that was usable by a nontechnical population. This issue was discussed at length in Chapter 9 on human factors, and involved the learning curve of one in the case of the ATMs. Some recent examples of rapid technological acclimation by the public includes the following examples:

Television
Direct long-distance dialing
Automatic automobile transmissions
Automatic teller machines
Video cassette recorders
Financial transaction cards
Mass transit cash card fare collection systems

Each of these technological changes shared important characteristics.

Easy and quick public training
Transparent technology—technology that did not need to be understood by the users
Quick understanding and acceptance of the new results and benefits
Effective marketing to motivate participation and use

The missing ingredient in these scenarios is the demonstrated massive public usage of information processing facilities. True, there are some exceptions, such as children in school or executives in some industries or some professional environments who have quickly learned to use information processing. But they are a small minority. If a method could be found to achieve a massive public usage of information processing, then the results would be to assist in overcoming some of our current but transient needs. Some believe that this massive usage would come with higher level languages. The Cobol, Fortran, and Basic languages helped to create a bigger reservoir of data processing users. These were the professionals.

They were motivated by employment or educational goals. But even these users had to master a set of skills to achieve their motivated goals: the language, the input and output, the diagnostic and debugging processes, the program changes, updating, the work station. Yes, the process was successful, but the users were a very special new generation of computer professionals.

Others believe that the very small computer is technology's answer to achieving massive public use of information processing. Although many personal computers have been sold, however mass use of information processing is not happening. After several years of intensive effort, small personal computers have reached perhaps 12–14 percent of U.S. households (as of year-end 1984). This compares with 80 percent of households with television after its first six years of mass marketing. Even now, a major vendor is selling ONLY 25 percent of its very small computers to households. If the annual industry sales are 8 million personal computer units and the 25 percent rate is typical, then the growth is 2 million households per year. At that rate it will take 45 to 50 years to fully saturate the United States home market. A personal computer selling rate to homes of ten times as fast would be needed to achieve a rapid home acceptance.

Is that a realistic expectation? What is required to use a very small computer? The languages, the key boards, the computer processes and procedures, the computer terminology and concepts, all contribute to maintaining the usage barrier.

MASSIVE USE OF INFORMATION PROCESSING

The Introduction started with the story of the waterwheel as a large central power plant. The migration to the fractional horsepower motor lead to a massive growth of motor use in our homes. It created the leisure time and productivity we enjoy today. But where are the motors? Most of the fractional horsepower motors are barely perceptible or are not even known to exist—the clock radio, the automatic thermostat, and the refrigerator are well-hidden examples. Yet we all use and enjoy the benefits of the small motor.

Those are the clues! Hidden, transparent, built-in to everyday operation and practices, but with added horsepower, results, and output. This is the format of the ultimate personal computer. Carry a pocketful! Put them into your telephone or television set. Put them into your information appliance. Tap the power of information processing for applications, for communications, and for data base access without any of the usual information processing complexities. No procedures, languages, complex keyboards, session starting, interactive details, or data processing education. The Smart Card opens an era that taps the power of information processing without the details or the specifics. Furthermore, the Smart Card is pocket portable, usable almost anywhere, and as convenient as the nearest Smart Card telephone or television set.

WHAT IS ULTIMATE?

To be an ultimate anything is not easy. It does not happen fast. It takes trial and error. It takes experience. It requires an agreement on the desired attributes. It takes the cooperation of many participants to make it occur. However, consider that 98 percent of United States households have television sets. That helps to make the television set an ultimate example of audio-visual communications and entertainment. What are some of its ultimate attributes?

Simplicity of operation
Consistency of operation
Predictability of execution
Transparency to technology
Repeatability between vendors

The television set is, however, a home body. It is generally not taken to the office or work place. To its attributes must be added those of the information and communications era.

Mobility
Portability
Transferability
Ease of self-diagnostics
Ease of repairability
Ease of recovery

WHAT IS THE ULTIMATE PERSONAL COMPUTER?

The ultimate personal computer is not a computer in the conventional sense. It is not a general purpose computer. It lacks any requirement for user understanding of languages, usage procedures, interconnected building blocks, or multifunction keyboards. The user knows nothing of operating systems, application programs, diagnostics, or recovery considerations. In short, the ultimate personal computer is one that requires *absolutely* no special training or knowledge.

The ultimate personal computer is totally mobile. The user can have an application and solution available by bringing one or several Smart Cards to any of the many devices that accept and interact with the Smart Cards. These may be any new telephone work station with Smart Card interface, or some new breed of information appliance whose operation has been tuned to the Smart Card approach. The menu displays and minimum keyboard units are designed to handle user Smart Cards and job cards and device cards. The configuration will vary, like the features of an automobile, and will be matched by the device card functions. When the application is limited to specific minimum configurations, the job card will recognize an inadequate configuration. These considerations will be given to the user in a manner consistent with normal Smart Card display interactions.

Tomorrow's ultimate personal computer is a hand-held package. The package has been prepared so that each one provides a specific application result. We will carry a set of these ultimate personal computers. There will be a library at home, in the office, and at our work location. Like a pocketbook, the cover will be immediately identifiable as to its application area and expected results. In more complex applications, several Smart Cards will be used. Each will provide a significant part of the results. Several may be used in a prescribed sequence to cover the full application structure.

The ultimate personal computer will be economically attractive. It will be the integrated circuit card publishing vehicle. It is tomorrow's application program delivery package. Smart cards will be published in large volumes, available on specialized racks at the corner supermarket. This will allow the purchaser to tap the economy of scale of mass production. The integrated circuit card price will continue to lower. Its storage capacity will grow to accommodate a greater range of application solutions. The upper end of the Smart Card spectrum will be the low end of the conventional personal computer. It will allow the experienced Smart Card user to move into general purpose computer facilities on a selective basis.

THE SMART CARD OPENS A MAJOR NEW APPROACH

The Smart Card combines the power of a microcomputer-on-a-chip with packaged application solutions using information processing fractional power. Each highly tailored Smart Card opens a new world of information processing for the unskilled. The Smart Card–based solutions are usable by those with relatively low or nonexistent information processing skills. The Smart Card requires very little, if any, information processing training. It is a simple procedural world, which taps the microprocessor productivity.

Central to this new approach are the Smart Card information protocols. They are critically important to the user, to the issuer, and to the ease of Smart Card operation and adaptation. The information protocols offer a uniform process that is quickly and easily mastered. It is the lever that opens the power of information processing to massive use.

THE SMART CARD SOCIETAL CONSEQUENCES

The United States Connection

Given the ease of Smart Card usage and adaptation what are the possible results? One is that the Smart Cards put information processing into the hands of the broad spectrum of a using population. This is independent of economic, experience, educational, or societal capabilities. If achieved, this would take broad portions of a disadvantaged population and give them access to information processing power. The Smart Card has the ability to provide information processing power with only the barest of educational preparation. Its information protocols handle the operational initiation details with little user assistance. Its tailored microcomputer program handles the application details with a minimum of user direction and interaction. When implemented, the Smart Card will offer ease of access to a broad set of useful application results. Since the user controls his Smart Card use, the user has direct control and protection of his private information. User-controlled privacy offers a significant achievement in this era of data base accessibility.

A most important Smart Card opportunity is the large users group that is already being conditioned in the marketplace today: the 200 million issued financial transaction cards, the 60,000 ATMs, and the

45 percent (and growing) of United States households that routinely use these capabilities. This opens a whole world of self-service. All locations, time periods, industries, and applications are legitimate expectations for Smart Card realization. The Smart Card will be an important tool in the hands of mankind. It will be a major public usage of chip technology.

The French Connection

In the chapter on field tests, the role and plan objectives of the French government and vendors were discussed. A key objective of the French is to achieve world leadership in the Smart Card. They want to create an exportable, advanced product set of terminals, system units, and Smart Cards.

This type of national goal setting, planning, and implementation is not new. Several important examples of success have resulted from similar efforts in Japan. The Japanese international export capability in small cars, high fidelity audio, video equipment, and cameras are but a few examples of worldwide technological leadership from a national planning start. Even now a Japanese collective effort to develop a fifth generation computer is being taken as a very serious objective and possibility.

The Japanese have succeeded in exploiting chip technology. They have not succeeded in any major way in the export of traditional information processing capability. The French Smart Card effort has now arrived on the scene with a nontraditional information processing capability. Their Smart Card proposal and action plan focuses on several important capabilities.

New system attachments to offset growing local communications expense.

New information protocols that allow a sharp expansion of access to bank credit while improving usage control.

New work station concepts to expand market access to financial transactions and services.

New Smart Card concepts that allow the general public to use remote information processing in a self-service mode, but without requiring information processing training.

The French connection is impressive. Perhaps they have invented a new information processing lever to overcome societal shortcomings.

If so, the French offerings will succeed in affecting our social needs at a basic public level. Is that good or bad? It may be good and helpful in solving a serious social need. It will be bad if it costs the United States in lost import funds and puts the United States in a second-class role in a major technological and vital world market. Fortunately, as the opportunity matrices show, there is much more to the Smart Card opportunity than the financial transaction Smart Card. The real challenge is in achieving the opportunities of the entire spectrum—while all the players are still close to the starting line.

17 Achieving Smart Card Acceptance

WHAT IS ACCEPTANCE?

There was a time in the card industry that a discussion and selection of card technology alternatives was simple. Describe the needs, list the specifications, check the economics, select the standards, set the timetable, and the implementation plan fell into place. Today the card media application environment is more sophisticated and more demanding. The marketplace competition, implementation economics, business gains and losses, transition considerations, and legal/tax changes exert great pressures and demands on the acceptance decision processes.

The variety of demands on acceptance has led to a series of important questions that demand answers. Consideration of these key questions and their answers is vital for preparing the acceptance case. These issues arise in industry meetings and discussions as well as preparing vital business proposals. Why do they arise? These seem to be motivated by inexperience and/or a drive to subjugate all decisions to pure cost-of-cards economics. It's important to confront these questions with specifics.

ISN'T THE TRACK 1 OPTION LESS EXPENSIVE THAN A SMART CARD SOLUTION? DOESN'T TRACK 1 GIVE MOST OF THE SAVINGS? AREN'T SMART CARDS MORE EXPENSIVE TO IMPLEMENT?

Facts: Magnetic stripes have been tested and are generally specified to a two-year product life by the card technology standards working groups.

On the average they have only a 15-month field life because they are given early expiration dates as a bad debt loss control feature. Smart Cards have been specified to a 36-month life by the card technology standards working groups. Smart Cards are expected to have an average 30-month field life, or twice that of the magnetic-striped card (track 1 or 2). The Smart Card superior dynamic updating and bad debt loss control, e.g., specific funds available as reduced by transaction value, for all transactions, on and off-line, allows designating a longer period to expiration.

The industry uses current magnetic-stripe card and issue costs at $.88 per card. With a switch to track 1, there needs to be added the security feature that is to prevent large-quantity reproduction of an issued card with track 1 content control fields. EMI/Malco's Watermark is offered for this facility. Unfortunately, there is no security technique of this type that has been reviewed and accepted by the international standards technology working groups. Reproducing the track 1 decision data base is valuable because it may specify that "no PIN is required" and a "minimum of on-line checking is needed." The track 1 card also requires more issue and data complexity. All of these costs are assumed to raise the per card costs to $.35 plus $1.00 for issue costs, or a total of $1.35 for 15 months average track 1 card field life by 1990 (see Table 12.6, page 145).

The magnetic stripe track 1 card implementation requires a new terminal in all accepting locations. The unit provides reading track 1 and logic for use of the new control features. The cost estimates are $100 purchase for the track 1 only interface reader. This does not allow for reading track 2 because the standard track 2 data is also found in track 1. However, there are thousands of in-lobby PIN capture units that record PIN validation fields in track 2 only. Hence, adding a track 2 reader is a most probable added expense.

The Smart Card, due to its in-card logic, can use a lower cost "Certificator" at low-volume locations. This less than $50 purchase unit validates PINs, captures transaction records in the Smart Card, captures the transaction amount, reduces the available funds balance in the Smart Card, and issues an authorization code from the Smart Card to be written on the imprinted forms set used to capture the transaction data. Depending on the number of cards issued as compared with the number of terminals to be installed (assumed to be 100:1) and the ratio of conventional-to-certificator installations (assumed to be a 40%:60% for Smart Cards), the relative costs are shown in the following table.

The Smart Card is assumed to offer a 12 percent reduction in bad debt losses based on a VISA study estimate in real time systems—which the Smart Card gets from its transaction amount capture. Track 1 does *not* have that function. The industry now has a $10 loss per issued card. In addition, the industry will take the bad debt loss reduction capability of the Smart Card and use it to increase market penetration. That is worth a $1000 loan balance times 50% of the new customers that use the revolving credit feature times 5% net income times a 4.8% added market penetration (based on the 12% loss reduction above).

All these facts come together in the following 1990 time frame comparison:

TABLE 17.1

ECONOMICS	STRIPE (T1)	SMART CARD	COMMENTS
Card Cost	$ .35	$2.00	Mastercard report
Issue	1.00	1.50	
2d 15 months	1.35	—	SC 30-month life
Terminal costs	1.00	.80	SC 60% at $65
Total	3.70	4.30	SC cost 16% more per card
Reduced credit loss (−)		(3.00)	12% × $10/yr × 2.5 yr
Added Mkt penetration		(3.00)	$1000 × 50% × 5%/yr × 4.8% × 2.5 yr
Net total	3.70	(1.70)	SC returns 146% of costs

Net: Track 1 is 16% less on a card-issued/terminal-required basis. Smart Card with credit loss reduction and added market penetration makes a 146% return for each issued card in 30 months.

DOESN'T THE SMART CARD REQUIRE MORE CAPITAL AND HAVE A GREATER RISK?

Several industry "leaders" keep referring to a billion dollar bill to switch to Smart Cards. That is, at best, a gross estimate. However, at this point it appears grossly overstated when considering the net costs of making the switch.

Facts: There is a basic cost of doing business incurred each year with the credit card business. It requires that 150 million magnetic-striped cards be issued every 15 months. It also requires that losses be paid for each year. These are the mandatory costs of staying in the credit card business. The terminals to support the new controls (track 1), need to be issued to keep losses from growing further. These are not investment risk issues. They are dollars spent to stay in business. The following chart looks at these costs for a 30-month transition cycle from magnetic-stripe (track 1) implementation to Smart Cards with a 150 million card base, in a 1990 time frame:

TABLE 17.2

ECONOMICS	STRIPE (T1)	SMART CARD	COMMENTS
1st 15 months	$225 Million	$364 Million	75 M Stripe; 75 M SC
2d 15 months	225 M	263 M	75 M SC
Losses, 2 yr	4,000 M	3,040 M	−6% yr 1; −12% yr 2; etc.
2M terminals	200 M	160 M	SC 60% at $65
Total	4,650 M	3,827 M	Minimum annual costs SC is 17% LESS
No SC Loss Reduction		960 M	
Total	4,650 M	4,787 M	Worst risk case SC 3%+
Include loss reduction & added market penetration		(2,850 M)	150 M cds × $7.60 × 2.5 yr
Total	4,650 M	977 M	Best Smart Card case SC is 79% *less* costly

Net: Smart Card implementation requires less capital on an annual outlay basis. Is there a risk that the Smart Card-based bad debt credit loss reduction will not materialize? The two alternatives have almost the same outlay. Furthermore, the risky but probable added Smart Card market penetration helps to reduce outlays.

DOES THE SMART CARD NEED NEW SERVICES TO BE ECONOMIC?

Facts: The Smart Card doubles expected card field life. The Smart Card internal logic allows use of lower-cost certificators. These are real savings with the current card-based services.

The prime economic impact of Smart Cards is bad debt loss reduction and the consequent added market penetration which that allows. Both of these added revenue opportunities are strictly based on current card-based services. The prior economic analyses were based on a 12 percent loss reduction figure from a key industry card servicer. That number may be understated. The servicer assessment made no attempt to capture and use the value of each transaction with the Smart Card. No attempt was made to use this running view of actual transaction values, on- and off-line, to cut bad debt losses before they are incurred.

The estimate of the potential Smart Card-based reduction of bad debt losses in the bank credit card might be as much as 50 percent. Furthermore, those savings could then be used to achieve an added market penetration of as much as 20 percent. That would put bank credit card penetration at 60 percent of the U.S. adults. That is the current level of retail credit card penetration.

Smart Card reduction of bad debt to 50% is worth an added $U.S. 3.80 per card per year ($10 × 50% − 1.20). Added market penetration by 20% is worth an added $3.80 per card per year [$1000 × 50% × 5% × (20% − 4.8%)]. Together these improvements offer $7.60 × 150 million cards per year or over $1.1 billion per year. That Smart Card return, alone, is more than twice the implementation costs.

Overall, without any consideration of new services, the Smart Card returns 146 percent of its issue costs on single-card life cycle basis (30 months). Overall, the Smart Card economics are appealing. One of the unfortunate facts is that the bank credit card industry does not have a generally available or acceptable set of models to test the economic impact of these issues offered.

The Smart Card international standards are being designed to provide up to 16 applications or services per card. This is by far more service capacity than has ever been available from an end user carried and used media. Thus the Smart Card will offer a very extensive capacity for new services. However, for initial acceptance consideration, that capacity should not be allowed to confuse the issue. The basic card and its current applications are economically justified now. Waiting for the development of the new services, or testing their acceptance before Smart Card use, will unduly delay realizing Smart Card savings for serious current problems, e.g., bad debt reduction and market penetration growth.

DOES THE SMART CARD INTRODUCE NEW BUSINESS RULES?

The largest loss category with bank credit cards is bad debt. It is over four times the fraud losses. Industry reports [*Nilson Report* (September 1986), Issue 388], based on first quarter 1986 losses, that the dollar losses for the year for ten major issuers will *double* over 1985 losses. The projected total bank credit card bad debt loss will be $U.S. 1.8 billion. That is 22 percent greater than 1985's $1.47 billion credit losses. That is a greater increase than the 18 percent growth in outstandings.

A recent industry servicer press report talked about a $U.S. 15 million reduction in card counterfeiting in the last year. At the same time credit losses were up $U.S. 330 million. That is a loss increase of more than 20 times the counterfeit reduction.

How do magnetic-striped bank credit cards control bad debt losses? *They don't!* Eighty percent of bad debt losses go through authorizing transactions that approve the transaction. How is bad debt controlled? Credit screening before card account granting and card issue is the first line of defense. This limits market penetration. The second alternative is to pick up the cards after they exceed their bad debt limit or force the card to expire. Hence, the average life of a bank credit card is only 15 months.

That's not very impressive control. Limit the market penetration. Limit the field life of the bank credit card. Or, extend either control and watch the losses increase—as is now happening from the loosening of credit screening last year in order to increase earnings.

Why doesn't the magnetic-striped card do a better job? Because the magnetic-striped bank credit card is *blind!* Three out of four transactions are off-line. A change in customer economics, account status, or the value and quantity of off-line use *cannot* be reflected in the stripe-contained data in POS usage. It only appears days later in the authorizing system for on-line transactions or weeks/months later in the bulletin. This is a major shortcoming of the track 1-contained control fields. The track 1 striped-based data cannot be updated after card issue based on the customer's current economic situation or actual off-line usage!

The Smart Card is electronically updated in EVERY TRANSACTION—on- and off-line. The Smart Card can capture three types of data: the transaction amount, the transaction mode (on- or off-line), and the transaction frequency, e.g., transaction per day, week, or other time period. This captured information is now examined with a NEW SET OF BUSINESS RULES. What are possible new business rules?

Number of consecutive transactions off-line prior to requiring on-line reconciliation. (One servicer is considering setting this at 10 uses or 30 days.) This control value may be combined with type or location of the accepting transaction point.

Maximum cumulative value of transactions requiring on-line reconciliation.

Available funds for the next 30 days of card usage.

Available credit line based on current payment record, for example:

Full credit line:	Not less than missed 1 payment in 3 months or made a timely 3d payment for a new card.
2/3d's credit line:	Missed 1 payment or 2d payment for a new card. Return to full line with 2 more on time payments.
1/3d credit line:	Missed 2 payments or 1st payment for a new card. Return to 2/3d's with more than 1 on-time payment.
0 credit line:	Missed 3 payments. Return to 1/3d with 1 payment.

Note. These rules also limit the buildup of available credit at initial card issue and use. On established accounts these rules start to reduce exposure at the first missed payment—not 60 or 90 days later.

Customer's economic or payment record change will be reflected in in-card controls at the next on-line reconciliation (not longer than 30 days).

The new captured data and the new business rules introduce an improved set of "before-the-fact" credit loss controls. That is, the captured data and rules offer improved and new controls:

New in-card controls:
Available balance, frequency of usage, and mode limits.

New off-line controls:
Available balance, value/frequency since last on-line usage and next on-line control date/period for off-line usage.

New on-line controls:
Monthly or account status review period. Control balances and activity controls changes or updates. Business rule control fields update. Reconciliation of off-line usage on on-line account changes. For example, writing a check against an entire transaction account with an issued Smart Card may be prohibited until the Smart Card is physically produced and the balances are checked.

WHAT SAVINGS WILL THESE BUSINESS RULES AND CONTROLS PRODUCE?

I guess a 50 percent reduction! That is an unevaluated guess. Where are the models that one may use to evaluate these new business rules?

These models are now being prepared and studied. The bank card industry has half a TRILLION dollars of card-based transactions in the United States. Annual losses are measured in BILLIONS and growing. Credit losses are approaching 2 percent of gross credit card sales and are a growing percentage. It is just a matter of time till these new risk management rules and models are understood and applied.

WILL THE SMART CARD BE A DEBIT OR A CREDIT CARD?

Consider current strategic directions. The debit or ATM access card is the strategic retail card. Strategic thrusts to interchange, POS, regionalization, self-service, high-productivity branches, remote- and off-premise banking and loan product access (overdraft, equity loans, and securities loans deliver funds to checking accounts). All of these product relations are building the dependence of the retail bank on the debit card. In fact, the U.S. banks have twice the dollar transaction value controlled through debit cards when compared to credit cards.

Why? Several good reasons. All bank customers are eligible for a debit card as compared to less than 40 percent of U.S. adults having access to a credit card. Debit cards are routinely handled with PIN's. Debit cards routinely are multiple account relations, e.g., savings and checking accounts. Debit cards emphasize the relationship with the issuing bank through the front of card logo. The added service relationships are easily shown on the reverse of the card.

What about the credit card? It is a payment device that accesses an unsecured loan. It is usually implemented in a different part of the bank than the retail branches. They may not coordinate PIN issue with the ATM debit card product managers. Customers will not use ANOTHER PIN than their ATM PIN.

Where does it all come together? The debit card is becoming both the universal retail access card and loan product card. It will be a small step to incorporate the credit function. Hence, it appears reasonable to think of the future Smart Card with its capacity for 16 different applications per card, as the combined debit and credit card of the future.

WHY DOES THE CARD HOLDER WANT A SMART CARD?

Actually most card holders really can't care less about the technology provided in the card. The customer wants service, convenience, speed, simplicity, safety, and, when possible, transaction mobility. That is, the ability to access the financial services at any convenient place. The other need, as the customer sees it, is that the equipment is working when the service is needed. Customers quickly learn to leave a checkbook at home when ATM services are provided.

There is another important reason why the card holder will want Smart Cards. The Smart Card will open up a number of major new card product offerings. For example, a credit card with a guaranteed monthly spending limit. A telephone credit card that is only usable in calling predesignated telephone numbers. A card that is usable in an emergency medical services facility to identify him and his medical insurance coverage. Few if any card holders will associate these new and important products with the Smart Card. But that is the technology which will allow leading financial service institutes to offer the next generation of media services to card holders.

SUMMARY

The focus of this acceptance review has been the financial transaction Smart Card. This is estimated by industry experts at 50 percent (by volume) of the future Smart Card market potential. The second largest category is cards associated with communications usage. These are estimated at 40 percent of the potential. However, the latter category may include both contact and contactless Smart Cards. These cards fit in the same economics as financial cards when used for credit card calls. The remaining Smart Cards include industrial, medical, government, and other high-value cards. On a Smart Card technology basis, those cards will be noncompatible with physically larger packages, additional content and performance, and, hence, higher value.

Acceptance will, of necessity, require hard specifics. These include economics, alternative technology assessment, and timeliness of action. At this junction, the Smart Card has tied itself to the growth and rise of the integrated circuit chip. The integrated circuit chip appears to be a technology with unlimited future. Hence, the Smart Card offers untold opportunities. The secret is to start now in facing the acceptance debate. If you are not involved at the forthcoming buildup point, you may be too late to participate in the main growth phase.

18 A Glossary of Terms

Acceptor: One who receives payment for a transaction with the use of a financial transaction card.

Access: Achieving physical entry or passage with the use of an appropriate medium (e.g., a badge or card). Achieving physical and/or electronic entry to a storage device and its content. See also *Access Card*.

Access Card: A human carried, machine-readable medium used to achieve physical entry or passage.

Account: A business relationship involving the exchange of funds, value, or credits.

Acquirer: The financial institution that receives payment information and/or media from an acceptor. See also *Acceptor*.

Active: An action results in response to outside simulation such as providing a response to the signals entered into a Smart Card. See also *Passive*.

Algorithm: A prescribed computational procedure such as an encryption process. See also *Data Encryption Algorithm*.

American National Standards Committee: The standards-setting body in America (ANSC, or also ANSI).

ANSC: See *American National Standards Committee.*

ANSI: See *American National Standards Committee.*

Anticrime: Features or functions intended to inhibit misuse or modification of payment media and systems, such as an alter-resistant signature panel.

Application: A commercial use or purpose. A method of applying or using a device such as a Smart Card used for a payment purpose.

ATM: See *Automatic Teller Machine.*

Audit: An examination of records or accounts to check their accuracy and status.

Authorization: A process of granting permission such as authorizing a card-based transaction by reference to the appropriate account and card usage records.

Authorization Code: A number issued and recorded to confirm that a valid authorization occurred.

Authorizer: Additional functions combined with a Smart Card in the same package to allow off-line control. It adds a keyboard, display, logic, and power to the package. It may share the Smart Card communications interface. It is used to perform logic and arithmetic with the Smart Card content, e.g., PIN entry validation and transaction amount acceptance within available funds.

Automatic Teller Machine: A self-service unit that allows the user, with a suitable identification and account relation, to carry on financial transactions such as a cash withdrawal (ATM).

Availability: Accessible for use when desired by the user.

Bank card: A financial transaction medium issued to customers by an appropriately chartered financial institution or Bank.

Bank Protection Act: A national legislative act that defines and establishes responsibilities within a bank for the bank's protection.

Bit: A unit of information having only one of two values, a zero or a one. The units are used in combination to express information such as characters or digits.

Black List: See *Hot List.*

Business Rules: The application logic and arithmetic applied to Smart Card contained control and activity information.

Byte: A combination of bits (usually 8 to 10) that defines the representation of a set of characters or symbols.

Card: A rectangular paper or plastic medium used to show information relating to its issuer, user, and acceptors. It may include appropriate control information such as the dates and services for which it is usable.

Card Acceptor: See *Acceptor.*

Card Authorization: See *Authorization*.

Card Issuer: See *Issuer*.

Card Number: See *Primary Account Number*.

Card Life Cycle: The sequence of steps from initial manufacturing to usage completion for a card.

Card Holder: The person or entity with whom an account relation is established and is signified by the issuance of a card.

Carte a Memoire: The French term for Smart Card.

Cash Dispenser: A self-service unit that dispenses currency. See also *Automatic Teller Machine*.

Cash Card: A payment transaction medium used to access funds set aside for direct payment transactions. The transaction does not require an identification of the account holder.

Certificator: A hand-held, stand-alone, device that provides off-line control. It has a keyboard, display, logic, power, and a Smart Card insertion and interface capability. It is used to perform logic and arithmetic with the Smart Card content, e.g., PIN entry validation and transaction amount acceptance within available funds.

Character: An alphabetic or numeric symbol.

Chip: A small square of thin, semiconductor material, such as silicon, that has been chemically processed to have a specific set of electrical characteristics such as circuits, storage, and/or logic elements.

Chip Card: See *Integrated Circuit Card*.

Clearing: The process of returning financial transactions from the acquirer financial institution to the issuer institution for reconciliation, billing, and statement use.

Contact: An electrical connecting surface between a Smart Card and its interfacing device that permits a flow of current.

Contactless: A connection between a Smart Card and its interfacing device that does not use a contact surface. In these devices a flow of current and signals is achieved by induction or high-frequency transmission techniques.

Credit: An amount of funds placed by a financial institution at the disposal of a customer, against which he may draw for payment transactions.

Credit Card: A card that signifies an established relationship between its issuing financial institution and the holder designated on the card.

Custom Chip: See *Tailored Chip*.

Data Capture: The recording of information for subsequent use and information processing.

Data Encryption Algorithm: An encryption process that is a United States national standard, an ANSC national standard and a financial industry standard. The process is a key-driven and reversible process (DEA).

Data Tag: A military dog tag or metallic identification device which has been changed to a plastic package of the same rectangular dimensions, but thicker, to which has been added Smart Card chip(s) and function.

DEA: See *Data Encryption Algorithm*.

Debit: A payment transaction against an account with deposited and available funds.

Debit Card: A card that signifies an established relationship between a financial institution and the card holder designated on the card. The card data may not distinguish between a credit or debit relationship.

Decryption: Converting scrambled information back to plain or clear text.

Descramble: See *Decryption*.

Dedicated Network: A communications facility established for a specific purpose, such as servicing point-of-sale facilities. Each remote terminal is assigned to a specific termination point.

Destroyable Link: A method of irreversibly removing or altering chip functional ability such as adding "value" increments by electrically altering a circuit by an action, such as burning out a fusible element or link.

Device Card: A Smart Card that carries information (logic and data), used to configure, operate and supervise a work station which accepts Smart Cards or contains Smart Card function.

Digit: A numeric symbol.

Digital Optical Laser Card: A portable medium that passively stores information in the form of high-density marks or bars.

Display Telephone: A public-dial network work station with voice input/output, destination identification digit entry, machine-readable card device, and a visual information presentation device for communicating with the user.

Disposable Card: A medium designed for a specific period or amount of use, such as the number of trips or telephone calls, after which the card no longer has any value and may be discarded.

Economic Justification: A monetary assessment of an application in which the costs and expenses of implementation and use are "acceptably" exceeded by the value of the end results as achieved by added or new results or cost avoidance.

EEPROM: Electronically erasable, programmable, read-only memory. This memory is electronically erasable and is nonvolatile.

Electron Card: A new VISA bank card that does not use embossing for designation of the holder name or account number.

Electronic Check Book: An INTAMIC designation for a Smart Card type of debit card transaction.

Electronic Directory: A French work station that uses display and keyboard facilities with on-line data base access to replace printed white and yellow page telephone directories.

Electronic Purse: An INTAMIC designation for a Smart Card cash card transaction. Transactions are probably cleared against the card issuer's account rather than the card holder's account.

Electronic Wallet: An INTAMIC designation for a Smart Card type of traveller's check, cleared like a debit card transaction.

Embossed: Raised characters on a plastic bank card. These are used for a quick data capture in an imprinting device to transfer the information from the raised characters to a forms set for later processing back to the card issuer and to provide a receipt to the card user.

Encryption: Converting clear or plain text to scrambled text with the use of a key-driven algorithm. See *Data Encryption Algorithm*.

EPROM: Electronically programmable, read-only memory. A semiconductor memory that is erasable with ultra-violet light. This is nonvolatile memory.

Exception: A transaction that does not receive authorization by the accepted rules, procedures, and account conditions. It must be rejected or additionally processed by other steps or rules.

Expiration Date: The time beyond which a card or account is not available for transaction use, unless an exception process is used to gain permission.

Fall Back Process: An alternative procedure to be followed when the primary action plan is not available or appears not to be working properly or securely.

Float: Funds that become available for account holder use prior to the completion of a related payment clearing from the acquirer to the issuer. This may occur on a funds deposit or withdrawal.

Floor Limit: A transaction value above which the transaction acceptor or acquirer must follow the payment guaranteeing financial institution authorizing rules or process.

Front End: The communications network interfacing equipment at a control point or central site.

Function: An assigned role or duty such as the functional capabilities of a Smart Card.

General Purpose Chip: A chip with electrical properties that are set for the handling of a common set of requirement, such as a microprocessor or storage unit.

Generic: Descriptive of an entire group or class.

Hand Geometry: A method of personal identification verification that uses selected dimensions, shape, and other pertinent characteristics of the card holder's hand.

Heat Dissipation: Energy which, when used in a chip, generates an elevated temperature of the exposed surfaces and must be transferred away in order to avoid chip, chip carrier, or card damage.

Hermetic: Completely sealed against external substances such as liquids and gasses.

Holder: The individual or business in whose name a payment device is issued. (This may not be the device user.)

Holograph: A method of producing an optically sensed image that has three dimensional characteristics.

Home Banking: Remote financial institution services and transactions through the use of communications and work stations in many locations including homes, businesses, and professional offices.

Hot List: A compilation of lost, stolen, or abused account cards, used as part of a financial transaction card payment transaction authorization. The list may be in printed or electronic form. See also *Positive List*. The list does not include unissued account numbers.

Host: The control or central point at which payment transaction and card usage statistics are captured for authorization, billings, and statements.

Human Factors: The design considerations that relate to efficient use of machines by human beings.

ICC: See *Integrated Circuit Card*.

ID Card Type 1: International standards designation for a financial transaction card, including the physical description and information content.

Increment Smart Card: A Smart Card whose application and function characteristics is found in the upper left corner of the Smart Card opportunity matrix. This card contains increments of value, supply, entrance or other units for issue and use.

InfoPro: See *Information Appliance*.

Information Appliance: A configuration of work station elements for Smart Card usage that requires no knowledge of computers for its use. Also *InfoPro*.

INTAMIC: See *International Association for Microcircuit Card*.

Integrated Circuit Card: The international standards designation for the Smart Card (ICC).

Interactive On-Line: An application in which there is dialogue between the user at a work station and a remote data base through a communications facility while the user is performing the application.

Interactive Smart Card Work Station: A work station with Smart Card application and product function as found in the lower right corner of the Smart Card opportunity matrix. It contains storage, logic, input and output, and communications facilities.

Interchange: An agreement between card-based service suppliers to accept each other's cards for a set of pre-established payment transactions.

Interchange Card: A medium designated for use in an interchange arrangement, usually indicated by an appropriate mark or sign on the reverse side of the card.

Interchange Smart Card: A Smart Card whose application and product function is found in the center of the Smart Card opportunity matrix. The financial transaction Smart Card is a typical example.

International Association for Microcircuit Card: A French-based international association of financial service institutions for the promotion of the Smart Card (INTAMIC).

ISO: See *International Standards Organization*.

International Standards Organization: The international standards-setting body (ISO).

Issuer: The financial institution responsible for providing a payment transaction account and its representative medium.

Job Card: A Smart Card that carries application-related information between a work station and its related financial institution, such as transaction data and hot lists for local card usage acceptance.

Journal: A listing of all pertinent payment transactions and the account(s) to which they apply.

K: Represents 1024 units.

Key: A personalizing value in the use of an encrypting algorithm similar to the use of a combination number for a vault. When the same value is used to control the encryption and decryption process it is called a *private key*. When a pair of different values is used to control a related encryption and decryption process it is called a *public key*. When a unique key is generated for a set of data it is called a *session key*.

Key Management: The process by which keys are distributed to usage points while kept in a protected form by encryption.

Learning Curve: A human factors phrase describing the change in user understanding of an activity over a period of increased usage experience. "Learning curve of one" indicates the user has reached a satisfactory level of performance in one use.

Limit: A specific quantity of usage or value available with an authorization process.

Lines of Defense: The steps available to respond to a security or operational threat.

Line Tapping: A criminal technique that involves connecting to conventional communications lines for listening to, changing, or introducing new information into a transaction between the end parties without their knowledge.

Liquid Crystal: An electrically driven display technology used for work stations that is small, lightweight, and with low power needs.

Logic: A sequence of instructions or steps for performing a specific job or task.

Logo: A name, symbol, or trademark for a company or institution.

Magnetic-Stripe Financial Transaction Card: A hand-carried, plastic medium for payment transactions. See also *Bank Card; ID Card Type 1*.

MasterCard: An international card servicer, authorization, and data capture facility used by financial institutions.

Microcircuit Card: The INTAMIC designation of an ICC type of financial transaction card.

Microprocessor: A microcomputer with all of its processing facilities on a single chip. Also called microprocessor-on-a-chip.

Multipurpose Card: A card intended for use in several types of financial payment applications such as debit, credit, and/or cash card functions.

Negative List: See *Hot List*.

NMOS: A moderate performance, moderate power requiring chip fabrication technology.

Nonvolatile Memory: A semiconductor memory that retains its content when power is removed.

OCR: See *Optical Character Recognition*.

Off-Line: A mode of work station operation in which the station is not connected to a control site for card-based authorization.

On-Line: A mode of work station operation in which the station is connected to a control site for card-based authorization.

Opportunity Matrix: See *Smart Card Opportunity Matrix*. This matrix defines the nine generic types of Smart Cards. They are a combination of application and product functions.

Optical Character Recognition: Character fonts that are machine-readable by optical techniques (OCR).

Open System Interconnection: An international standard for describing the interaction of computer systems through communications link characteristics by allocating the information functions into seven distinct layers (OSI).

OSI: See *Open System Interconnection.*

Package: A physical container, case, or enclosure for the integrated circuit chip(s) in a Smart Card.

PAN: See *Primary Account Number*.

Passive: Subject to action without a responding action. For example, reading the content of a magnetic stripe. See also *Active*.

Payment Transaction: The exchange of value for goods or services.

PC: See *Personal Computer*.

Personal Computer: A small computer with a selling price up to $10,000 and containing a display, keyboard, processor and data storage elements (PC).

Personal Identification Number: A four- to twelve-character field entered by a card user to validate that they are the proper individual to use the card (PIN).

Personal Identification Verification: Techniques used to test physical traits to validate an individual's unique characteristics (PIV).

Personalization: The process of entering information into a Smart Card that ties it uniquely to a given holder and account, including an appropriate PIN acceptance value.

PIN: See *Personal Identification Number*.

PIN Pad: A keypad for entering PIN values at a work station.

PIV: See *Personal Identification Verification*.

Plasticizers: Chemical used in the materials and fabrication of plastic cards. They may emit fumes for extended periods.

Point-of-Sale: The location at which payment transactions occur for the exchange of value for goods or services. Also, the designation of work stations used to implement these transactions.

Portable Work Stations: A work station with a weight low enough to be considered easily carried (e.g., up to 30 pounds). See also *Work Station*.

Positive List: A description of account numbers, available balance, and other pertinent data for all active and issued account numbers. Some of these may also appear on a hot list.

Prepaid Card: See *Cash Card*.

Prepaid Telephone Payment Card: A cash card that is purchased in telephone call value increments.

Primary Account Number: The primary account number that is the international standard designation for the account identification with financial transaction cards (PAN).

Private Key: See *Key*.

Proprietary Card: A card with the issuer's logo or identification that is intended for use with the facilities and equipment of the issuer institution (e.g., an ATM access card). More recently, these cards are also accepted in shared and interchange facilities.

Protocol: A specified procedure or process used to achieve a specific and common result, such as a network communications message format and content agreement.

Public Key: See *Key*.

Record: To capture information in a form that allows it to be subsequently read or to which logic may be referred.

Registration Authority: The ISO ID Card Type 1 standards designated group for issuing and assigning the industry and issuer portion of the PAN.

Remote: A systems attachment mode for Smart Card-based telephone type work stations that allow payment transactions to be accomplished through the public-switched (dial) network.

Retinal Scan: A PIV technique based on an infrared scan of the eye retina.

Runaway card: A loss or stolen financial transaction card that is being misused frequently and fast.

Secure Card Properties: Built-in card characteristics that logically tie the magnetic stripe and, possibly, its content to the specific physical card.

Security: Measures taken to achieve a reasonable freedom from criminal, fraudulent, and vandalizing actions while maintaining sensitivity to unexpected attacks or system failures that cannot be distinguished from attacks.

Signature Dynamics: A PIV technique based on pen acceleration and pressure measurements during signature writing.

Signature Panel: A plastic card surface used to capture and display the card holder's signature.

Shared System: A group of payment or self-service facilities in which several financial institutions have agreed to acceptance of the cards and transactions of holders from each issuer.

Skimming: A criminal technique that copies magnetic-stripe content of financial transaction cards by placing a stripe reader near a conventional point-of-sale facility.

Smart Card: See *Integrated Circuit Card*.

Smart Card Opportunity Matrix: See *Opportunity Matrix*.

Special Purpose Chip Design: See *Tailored Chip*.

Standalone: A method of work station operation that relies only on self-contained data and instructions to achieve payment transaction authorization and transaction completion. Captured payment transaction data must be forwarded for subsequent billing, statements, and reconciliation.

Standard: A voluntary agreement to a uniform and consistent method or data presentation used to achieve a common action or result.

Storage: An electronic and/or mechanical-magnetic device that holds information for subsequent use or retrieval.

Store and Forward: A system attachment mode for Smart Card-based work stations that allows routine payment transactions to be accomplished at a standalone work station with the pertinent data to be captured for later batch transfer (physical or electronic) to the acquiring financial institution.

Stripe: A magnetic recording material band on a plastic card used for financial transactions.

Super Smart Card: A VISA proposed Smart Card that combines a Smart Card and authorizer in the same physical package.

Supplier: A manufacturer of plastic cards.

Survivability: Built-in features, functions, and characteristics that help to assure that a Smart Card will last through its intended life.

T and E Card: A financial transaction card issued by a travel and entertainment card company such as American Express.

Tailored Chip: A chip whose characteristics have been set for the handling of a specific set of application or job requirements.

Telephone Card: A card issued by a telephone company for the payment of telephone calls through the usual telephone service billing process.

Terminal: See *Work Station*.

T1: Track one—the high-density, alphanumeric content, magnetic-stripe recording established for on-line airline industry usage.

T2: Track two—the low-density, numeric content, magnetic-stripe recording established for financial transaction usage.

T3: Track three—the high-density, numeric content, magnetic-stripe established for rerecording data base for usage in the savings and loan/banking industry.

Transaction: A business or payment event for the exchange of value for goods or services.

Transistor: A discrete electronic component whose use preceded chip-based devices.

Two-Party War Game: A term descriptive of the continuing need to be sensitive and responsive to security breaches.

Underwriters: A nonprofit American testing agency for the insurance industry.

Universal Product Code: A bar code used for product identification in supermarkets. The code is sensed by laser scanners (UPC).

UPC: See *Universal Product Code*.

Updatable: A storage medium that allows for rerecording data so as to allow retention of the latest form of a data field(s).

User: The person presenting a payment card for a financial transaction. The user may or may not be the card holder to whom the card was issued.

User's Card: A card presented by a user. See also *User*.

Validate: To substantiate that the elements of a payment financial transaction are correct, appropriate, and acceptable.

Vendor: A manufacturer of data-processing products.

Videotex: A remote television service operating with a telephone line-distributed bit stream that is converted to a graphic display frame(s) at the TV set. Remote input is communicated through the telephone line.

VISA: An international card servicer, authorization, and data capture facility used by financial institutions.

Vital Data: Very important information that requires encryption protection, such as an encrypted PIN.

Work Station: A combination of input and output devices intended to provide transaction implementation.

Write-Once: A storage medium intended to receive information in the additional storage area.

Zones: Areas of Smart Card storage designated for free access, limited access, or no access.

Index